Cities and Development

By 2030 more than 60 per cent of the world's population will live in urban areas, with most of the world's population growth over the next 25 years being absorbed by cities and towns in low- and middle-income countries. What are the consequences of this shift? Demographic pressure already strains the capacity of local and national governments to manage urban change. Today, nearly 1 billion people live in slums, and in the absence of significant intervention that number is set to double in the next two decades. Will our future be dominated by mega-cities of poverty and despair, or can urbanisation be harnessed to advance human and economic development?

Cities and Development provides a critical exploration of the dynamic relationship between urbanism and development. Highlighting both the challenges and opportunities associated with rapid urban change, the book surveys: the historical relationship between urbanisation and development; the role cities play in fostering economic growth in a globalising world; the unique characteristics of urban poverty and the poor record of interventions designed to tackle it; the complexities of managing urban environments; issues of urban crime, violence, war and terrorism in contemporary cities; and the importance of urban planning, governance and politics in shaping city futures.

This book brings into conversation debates from urban and development studies and highlights the strengths and weakness of current policy and planning responses to the contemporary urban challenge. It includes research-oriented supplements in the form of summaries, boxed case studies, development questions and further reading. The book is intended for senior undergraduate and graduate students interested in urban, international and development studies, as well as policy makers and planners concerned with equitable and sustainable urban development.

Jo Beall is Professor of Development Studies in the Development Studies Institute (DESTIN) and a research programme director at the Crisis States Research Centre (CSRC), both at the London School of Economics (LSE). She has extensive research experience in Africa and Asia, and has consulted for a wide range of international development agencies including the OECD, UN-Habitat, UNDP and World Bank. Her other books include *Funding Local Governance: Small Grants for Democracy and Development*, *Uniting a Divided City: Governance and Social Exclusion in Johannesburg*, and *A City For All: Valuing Difference and Working with Diversity.*

Sean Fox is a Research Associate at the CSRC and a PhD candidate in the Development Studies Institute at the London School of Economics (LSE). He has worked as a consultant on urban poverty and development issues for several non-profit organisations. He holds a BA in Economics from the University of California at Santa Cruz and an MSc in Development Management from the LSE.

Routledge Perspectives on Development

Series Editor: Professor Tony Binns, *University of Otago*

The *Perspectives on Development* series will provide an invaluable, up-to-date and refreshing approach to key development issues for academics and students working in the field of development, in disciplines such as anthropology, economics, geography, international relations, politics and sociology. The series will also be of particular interest to those working in interdisciplinary fields, such as area studies (African, Asian and Latin American Studies), development studies, rural and urban studies, travel and tourism.

If you would like to submit a book proposal for the series, please contact Tony Binns on j.a.binns@geography.otago.ac.nz.

Published:

David W. Drakakis-Smith
Third World Cities, 2nd edition

Kenneth Lynch
Rural-Urban Interactions in the Developing World

Nicola Ansell
Children, Youth and Development

Katie Willis
Theories and Practices of Development

Jennifer A. Elliott
An Introduction to Sustainable Development, 3rd edition

Chris Barrow
Environmental Management and Development

Janet Henshall Momsen
Gender and Development

Richard Sharpley and David J. Telfer
Tourism and Development

Andrew McGregor
Southeast Asian Development

W.T.S. Gould
Population and Development

Cheryl McEwan
Postcolonialism and Development

Roger Mac Ginty and Andrew Williams
Conflict and Development

Andrew Collins
Disaster and Development

David Lewis and Nazneen Kanji
Non-Governmental Organisations and Development

Jo Beall and Sean Fox
Cities and Development

Forthcoming:

Janet Henshall Momsen
Gender and Development, 2nd edition

Clive Agnew and Philip Woodhouse
Water Resources and Development

David Hudson
Global Finance and Development

Michael Tribe, Frederick Nixon and Andrew Sumner
Economics and Development Studies

Tony Binns and Alan Dixon
Africa: Diversity and Development

Tony Binns, Christo Fabricius and Etienne Nel
Local Knowledge, Environment and Development

Andrea Cornwall
Participation and Development

Heather Marquette
Politics and Development

E.M. Young
Food and Development

Hazel Barrett
Health and Development

Cities and Development

Jo Beall and Sean Fox

LONDON AND NEW YORK

First published 2009 by Routledge
2 Park Square, Milton Park, Abingdon, Oxon, OX14 4RN

Simultaneously published in the USA and Canada
by Routledge
270 Madison Avenue, New York, NY 10016

Routledge is an imprint of the Taylor & Francis Group

Typeset in Times New Roman and Franklin Gothic by
Keystroke, 28 High Street, Tettenhall, Wolverhampton
Printed and bound in Great Britain by
CPI Antony Rowe, Chippenham, Wiltshire

British Library Cataloguing in Publication Data
A catalogue record for this book is available from the British Library

Library of Congress Cataloguing in Publication Data
Beall, Jo, 1952–
Cities and development / Jo Beall and Sean Fox.
p. cm.
Includes bibliographical references and index.
1. Urbanization—Economic aspects. 2. Urbanization—History—21st century.
3. Cities and towns—Growth—History—21st century. 4. Urban economics—History—21st century. 5. Poverty—History—21st century. I. Fox, Sean, 1979– II. Title.
HT371.B43 2009
307.76—dc22
2008052776

ISBN 13: 978–0–415–39098–9 (hbk)
ISBN 13: 978–0–415–39099–6 (pbk)
ISBN 13: 978–0–203–08645–2 (ebk)

ISBN 10: 0–415–39098–2 (hbk)
ISBN 10: 0–415–39099–0 (pbk)
ISBN 10: 0–203–08645–7 (ebk)

This book is dedicated to Charles Tilly and Jane Jacobs, both great urban scholars, who passed away during the writing of this book, but who influenced it profoundly.

Contents

Plates

Figures

Tables

Boxes

Acknowledgements

We would like to acknowledge the many individuals and organisations that helped make this book possible. In particular we would like to thank Stuart Corbridge, Head of Institute, and all our colleagues in the Development Studies Institute (DESTIN) at the London School of Economics (LSE) for providing a stimulating and challenging intellectual environment. The Crisis States Research Centre (CSRC) supported this project both materially and intellectually, and we are grateful to the Director, James Putzel and our fellow researchers at the Centre who supplied thought-provoking seminars and dialogues.

There is a vibrant urban studies community at the LSE that has informed our thinking. We are particularly indebted to those who share our interest in the relationship between urban and development studies, including Sharad Chari, Sylvia Chant, Lucy Earle, Daniel Esser, Gareth Jones, Sunil Kumar, Kate Meagher, and David Satterthwaite.

Jo wishes to thank a special cohort of family and friends without whose received wisdoms and consummate hospitality her contribution would have been all the more limited: Paul, Brit and Marg Adams, Anna Arthur, Andie Ball, Charlie Beall, Rodney Bolt, Owen Crankshaw, Deborah James, Barbara Harriss-White, Shireen Hassim, Naila Kabeer, Nazneen Kanji, David Lewis, Francie Lund, Shula Marks, Chris and Norma McCormack, Sue Parnell, Patrick Pearson, Ann Perry, Edgar Pieterse, Jenny Robinson, Harold and Sonja Roffey, Ari Sitas, Anthony Swift, Mirjam van Donk and Astrid von Kotze.

Sean would like to thank Alan Mabin and Margot Rubin for generous hospitality and intellectual inspiration at CUBES, where a significant portion of the final manuscript was prepared, and acknowledge the love and support of Susan Fox, Robert Fox and Estefania Jover.

Jointly we wish to recognise the administrative support afforded us at the LSE by Drucilla Daley, Stephanie Davies, Wendy Foulds and Sue Redgrave. We are grateful for the good humour and gentle persistence of Michael Jones, as well as the heroic patience of Andrew Mould at Routledge. We would especially like to thank series editor Tony Binns for his guidance, and the instructive feedback on an earlier draft from two exacting but crucially perceptive anonymous reviewers.

Between us we have two very important final debts to acknowledge: Dennis Rodgers for good ideas, good taste and good laughs; and Tom Goodfellow for invaluable research assistance, a sharp editorial eye, and crisp scholarly critique.

Abbreviations

AD	Anno Domini
AGETIP	Agence d'Exécution des Travaux d'Intérêt Public
AIDS	Acquired Immune Deficiency Syndrome
ASH	Aided Self-Help
BBC	British Broadcasting Corporation
BCE	Before the Common Era
BOT	build-own-transfer
BRT	bus-based rapid transit
CEPAL	Economic Commission for Latin America
COHRE	Centre on Housing Rights and Evictions
CSIR	Council for Scientific and Industrial Research (South Africa)
DFID	Department for International Development (UK)
GATT	General Agreement on Tariffs and Trade
GaWC	global and world cities
GDP	gross domestic product
HDI	Human Development Index
HDR	*Human Development Report*
HIV	Human Immunodeficiency Virus
ICWE	International Conference on Water and the Environment
IDP	integrated development plans (South Africa)
IDPs	internally displaced persons
IFC	International Finance Corporation (branch of World Bank investing in private sector)
IFI	international financial institution
ILO	International Labour Organization
IMF	International Monetary Fund
IR	Indian Railways
IT	information technology
LAC	Latin America and Caribbean (region)

MDGs	Millennium Development Goals
MHP	Million Houses Programme (Sri Lanka)
MNC	multinational corporation
MUDH	Ministry of Urban Development and Housing (Afghanistan)
NGO	non-governmental organisation
NHC	National Housing Corporation (Tanzania)
NPM	new public management
NPR	National Public Radio (United States)
NSDF	National Slum Dwellers' Federation (India)
OECD	Organization for Economic Cooperation and Development
OSN	Obras Sanitarias de la Nacion (water and sanitation authority for Buenos Aires, Argentina)
PAMSCAD	Programme of Action to Mitigate the Social Cost of Adjustment (Ghana)
PPP	public–private partnership
PRSP	Poverty Reduction Strategy Paper
SAP	Structural Adjustment Programme
SDI	Shack/Slum Dwellers International
SEWA	Self-Employed Women's Association (India)
SIDA	Swedish International Development Cooperation Agency
SLF	sustainable livelihoods framework
SPARC	Society for the Promotion of Area Resource Centres (India)
SWM	solid waste management
TNC	transnational corporation
UBT	urban bias thesis
UNCHS	United Nations Centre for Human Settlements (former name for UN-Habitat)
UNDP	United Nations Development Programme
UN-Habitat	United Nations agency dedicated to urban development issues
UNHCR	United Nations High Commission for Refugees
UNICEF	United Nations Children's Fund
USCG	Urban Sector Consultative Group (Kabul, Afghanistan)
VoP	Voices of the Poor (project)
WCED	World Commission on Environment and Development (Brundtland Report)
WDR	*World Development Report*
WHO	World Health Organization
WTO	World Trade Organization

1 Development in the first urban century

Introduction

This book provides an introduction to some of the key issues surrounding the challenges of urban poverty and development in the twenty-first century. It comes at a critical moment. For the first time in history more people around the world live in urban areas than in rural ones. All economically advanced nations are predominantly urban societies, having urbanised rapidly during the course of the nineteenth and twentieth centuries. But most low- and middle-income countries remained (until relatively recently) predominantly rural. This is changing fast. Cities of unprecedented size are emerging; slums, shanty-towns and squatter settlements are growing; urban poverty is rising; war and terrorism threaten the security of urban dwellers in rich and poor countries alike; and the stability of the global environment is under threat, with particular implications for urban centres. At the same time, cities have historically played an important role as drivers of social, political and economic transformations; they are social melting pots, nodes of regional and international communication and transportation, engines of economic growth, seats of political power and iconic cultural spaces. As such, the relationship between cities and development is complex.

In this book we provide a global and interdisciplinary perspective on the relationship between cities and development. Throughout the book, we refer to those regions and countries that have traditionally been referred to as the 'Third World', 'developing' and the 'global South' as 'low and middle income'. We recognise the difficulties associated with using a collective descriptor like 'low and middle income', not least because it obscures the many and considerable differences within and between countries across Africa, Asia, Latin America and the Caribbean. Nevertheless, it suggests a more disaggregated analysis than homogenising terms such as the global South, or the now anachronistic Third World. The term 'low- and middle-income countries' also suffers from the fact that it primarily refers to differences in levels of economic development, whereas in this book we are concerned as much with political and social change. Nevertheless, we have chosen this nomenclature over the more normative distinction between 'developing' and 'developed' countries, which implies not only set development hierarchies but also prescriptive end goals. Drawing on urban and development theory, we explore the relationship between cities and development, broadly defined. We address the contributions that cities have made to development processes, as well as the unique economic, political and social development challenges posed by complex urban environments. We begin, however, by defining our terms.

Defining 'cities' and 'development'

What is a city? Most people recognise a city when they see one, but when it comes to providing a concise definition things get a bit complicated. There are conflicting technical-administrative definitions of a 'city' versus an 'urban area' versus a 'suburb' versus a 'town' and so on. For census purposes, cities are variously defined by their administrative status, population size, 'urban characteristic' or economic function. Of the 228 countries tracked by the United Nations (UN), 36 per cent use strictly administrative criteria for classifying cities, 25 per cent use population size, and 11 per cent have no definitive criteria (United Nations 2004: 104).

For our purposes, technical definitions are of interest only in so far as they affect estimates of urban populations. For analytical purposes the more qualitative definitions provided by urban scholars are of greater interest. These definitions focus on the social, economic and political processes that manifest in cities, and the link between these processes and the construction, organisation and use of space. Chicago School sociologist Lewis Wirth, whose essay 'Urbanism as a way of life' is a classic in the study of urbanism, introduced the most basic and enduring definition of a city. Wirth defined a city as a 'relatively large, dense, and permanent settlement of socially heterogeneous individuals' (Wirth 1938: 8). Importantly, he argued that these conditions – size, density and heterogeneity – create a distinctly 'urban way of life' and an identifiable 'urban personality'. It is the unique nature of the social, political, economic and cultural life of cities – or *urbanism* – that lies at the heart of urban scholarship.

Similarly, Lewis Mumford, one of the great urban scholars of the twentieth century, offered the following definition of a city:

> The essential physical means of a city's existence are the fixed site, the durable shelter, the permanent facilities for assembly, interchange, and storage; the essential social means are the social division of labour, which serves not merely the economic life but the cultural processes. The city in its complete sense, then, is a geographical plexus, an economic organisation, an institutional process, a theatre of social action, and an aesthetic symbol of collective unity.
>
> (Mumford 1937: 93–94)

This definition highlights the spatial dynamics of a built environment that serves as a 'theatre' of human interactions as well as a reflection of social relations. But Mumford is also attentive to the fundamental influences of size and density, for as Mumford observes:

> without the social drama that comes into existence through the focusing and intensification of group activity there is not a single function performed in the city that cannot be performed – and has not in fact been performed – in the open country.
>
> (Mumford 1937: 94)

It is the concentration – or congregation – of human energies and activities that brings a space to life and gives it a distinctly urban character (see Plate 1). Many authors before and since have sought to define or redefine the city, but the definitions offered by Wirth and Mumford remain fundamental to our conceptualisation of cities.

These definitions of urbanism are clearly universal – they are defining characteristics of urban agglomerations over time and everywhere. But any discussion of cities and development has to address this question: what distinguishes cities in the low- and middle-income countries of Africa, Asia or Latin America from those in the high-income countries of Europe, Japan or North America? Why write a book about cities and development? Despite the diverse historical, social, political, demographic and economic experiences of cities in low- and

Plate 1 Commuters in Shanghai

Source: Tom Goodfellow

middle-income countries across the world, they often share similar challenges, such as high unemployment, poor housing and infrastructure, inadequate services, dangerous environments, malnutrition and ill-health (Drakakis-Smith 2000; Gilbert and Gugler 1992; Gugler 1988). It is precisely these kinds of challenges with which development scholars and professionals are concerned.

What is *development*? The very word development implies change, and indeed those interested in development (as scholars, professionals and members of the general public) are generally concerned with understanding how and why communities and societies change in particular ways, and how change can be consciously catalysed or directed to serve certain goals. There are essentially two dimensions of change that motivate development research and practice: progress and differentiation. Progress implies positive change; differentiation refers to the unevenness of progress across time and space. For example, one of the most enduring questions in the field of development is: why and how have some countries become rich (i.e. progress) while others have remained persistently poor (i.e. differentiation)? Such questions raise many issues that are hotly debated in the field. How do we define and measure progress? Is material wealth the ultimate goal? What about liberty, leisure or environmental sustainability? What aspects of differentiation are most pressing (e.g. inequality within or between countries) and what can or should be done about it? Can progress (however defined) be achieved without differentiation?

These are just a handful of the difficult questions that motivate scholars in *development studies*, which is a problem-driven, interdisciplinary field of study incorporating theoretical insights and methodological tools from anthropology, demography, economics, geography, history, political science and sociology (among others). Development scholars draw on these diverse disciplines in order to understand why and how progress and differentiation within and between human populations occur. Through their research, development scholars contribute to the *practice of development*, informing those who 'do' development in some capacity. The practice of development refers to conscious interventions designed to enhance the quality of people's lives in some way. Who 'does' development? Development actors may include individuals, communities, governments, bilateral agencies (such as USAID and the United Kingdom's Department for International Development (DFID)), multilateral agencies (such as the United Nations Development Programme (UNDP), World Bank and International Monetary Fund (IMF)) and non-governmental organisations (NGOs), such as Oxfam, CARE International, ActionAid, Christian Aid and Save the Children. These actors, sometimes in collaboration and sometimes with conflicting objectives, are involved in the design and implementation of policies at various scales (i.e. from local to global), investments in the economy, public goods provision, advocacy and awareness-raising efforts around particular issues, the mobilisation of communities or interest groups, administration and governance, and so on.

Thus development is fundamentally about change, and in particular progress and differentiation; development studies is an academic field of enquiry that seeks to define, critique and understand progress and differentiation; and those engaged in the practice of development seek to affect change through public action.

There is, however, an inescapably normative nature to both the study and the practice of development. What do we really mean by development and how do we define progress and differentiation? While definitions of progress are forever subject to debate, we adopt a definition of development advanced by Nobel Prize winning economist Amartya Sen (1999) that provides a framework for thinking about and evaluating progress and differentiation, and hence development. In his book,

Development as Freedom, Sen (1999: 36) argues that freedom is both the primary end and principal means of development. Freedom, in Sen's formulation, means being able to live the kind of life that you value; to be able to choose *how* to live your life. According to Sen:

> Development requires the removal of major sources of unfreedom: poverty as well as tyranny, poor economic opportunities as well as systematic deprivation, neglect of public facilities as well as intolerance or overactivity of repressive states.
>
> (Sen 1999: 3)

From this perspective, which emphasises *human development*, progress can be understood as the expansion of real, substantive freedoms to human populations across the globe and differentiation to the observable unevenness of freedom across populations within and between countries. To this powerful definition of development as freedom, however, we must add a second dimension: sustainability. The expansion of industrial economies, which has allowed some to overcome basic unfreedoms related to material well-being, was facilitated by the rapid expansion of natural resource extraction, and to the use of non-renewable energy resources and emissions-generating technologies. While industrialisation has undoubtedly expanded real freedoms for millions around the world, the process of industrialisation has created environmental damage that now threatens present and future generations. *Sustainable* development, then, can be understood as the process of expanding real freedoms through means that do not threaten the possibility of future generations to enjoy and continue expanding real freedoms. Climate change is a direct threat to freedom, as it entails risks and costs that generate uncertainty and demand the diversion of resources away from satisfying other wants and needs. Expanding real freedoms today and in the future therefore requires environmentally sustainable strategies for satisfying our material needs and wants.

Having defined our terms (see Box 1.1), we are left with the motivating question for this text: what is the relationship between cities and development? In exploring this question, we are concerned with not only progress and differentiation within and between cities, but also the broader relationship between the social, economic and political processes that manifest in cities and the sustainable expansion of real freedoms within and across nations. The remainder of this chapter is devoted to establishing the context for our investigation. The following section provides an overview of key trends in development theory, policy and practice; the subsequent section examines how and why the relationship

Box 1.1

Key definitions in urban and development studies

City

A relatively large, dense and permanent human settlement. Cities are also referred to as towns, urban agglomerations or urban settlements. There is no standard set of criteria for identifying a settlement as urban, although population size, economic function and political/administrative status are frequently employed.

Urbanism

The way of life unique to habitation in a city. Generally speaking, urbanism refers to the unique social, cultural, economic and political dynamics that arise in densely populated human settlements. Importantly, urbanism is not necessarily confined to cities. Some social scientists have argued that urban ways of living, although endemic to cities, are exported to rural areas – particularly in advanced economies. For example, urban culture and consumption patterns are frequently found in rural and urban areas alike.

Development

The word 'development' takes on different meanings depending on context. Most often, the word is used to denote positive change – or *progress*. Progress can be defined in many ways. For the purposes of this book, development is equated with the expansion of substantive freedoms to individuals and communities across the globe. For Amartya Sen (whose definition of development we employ), freedom is both instrumental to achieving development and is the ultimate goal of development.

Development studies

An academic field of enquiry that revolves around theoretical and empirical studies of progress and differentiation – or the unevenness of progress across time and space. Development studies is characterised by its interdisciplinary nature, drawing on insights and methods from many social sciences, including demography, economics, geography, history, political science and sociology (among others).

Practice of development

As a practical activity, development implies conscious interventions designed to enhance the quality of people's lives in some way. Development actors may include individuals, communities, governments, aid agencies, civil society organisations and non-governmental organisations.

Sustainable development

The expansion of real freedoms through means that do not threaten the possibility of future generations to enjoy and continue expanding real freedoms. In particular, sustainable development refers to interventions that expand real freedoms while taking account of long-term environmental consequences. Sustainability can be assessed both at local and global levels.

between cities and development has remained somewhat peripheral in development studies. Finally, we review key demographic trends that underscore the importance of tackling this subject, and conclude by outlining the remaining chapters.

Key trends in development theory, policy and practice

The concept of development, and contemporary debates surrounding it, can be traced back at least to the Age of Enlightenment and the emergence of industrial capitalism in Europe in the eighteenth and nineteenth centuries. Adam Smith, in his *Enquiry into the Nature and Causes of the Wealth of Nations* (1776), famously celebrated the productivity gains that could be realised through the division of labour in production, and at the work of the 'invisible hand' of the market, which channels individual self-interest into an orderly system of production and exchange. He was one of the early proponents of trade liberalisation, which he saw as necessary to unleash the great potential of market forces. However, by the nineteenth century it had become clear that industrial capitalism had a dark side.

In 1842 the German social scientist Friedrich Engels was sent by his father to work in a Manchester textile firm. It was here that he observed the dreadful conditions under which English mill workers laboured, conditions he subsequently exposed and questioned along with Karl Marx. Marx, of course, went on to write some of the most influential and enduring critiques of capitalism and the social dislocations associated with industrialisation, and in doing so he posited a theory of history influenced by the work of German philosopher Georg Hegel. Hegel had theorised history as progress: a 'progression to the better' in the realm of metaphysics and the human spirit. Marx famously turned Hegel on his head and concentrated his formidable analytic prowess on the nature of progress in the material world (Leys 1996). He theorised history as a progression through various modes of production – communal, feudal, capitalist and eventually communist/socialist – with each step along the way representing progress towards the full realisation of human potential. Capitalism, according to Marx, would give way to socialism or communism due to the inherent contradictions of the system, which was clearly generating great wealth and great inequality simultaneously and on a historically unprecedented scale. According to Marx and Engels (1848):

> The development of Modern Industry, therefore, cuts from under its feet the very foundation on which the bourgeoisie produces and appropriates products. What the bourgeoisie therefore produces, above all, are its own grave-diggers. Its fall and the victory of the proletariat are equally inevitable.
>
> (Marx and Engels 1848)

This materialist theory of history echoed the dialectical logic of Hegel's philosophy, but also the evolutionary logic of another influential thinker of the nineteenth century – Charles Darwin. Taken together, the intellectual legacy of these (and other) Enlightenment thinkers was an evolutionary understanding of social change. The emphasis that these early observers of social processes placed on notions of evolution and progress, as well as the human consequences of socio-economic change (i.e. differentiation), formed the basis for subsequent debates concerning development theory and practice.

Despite eighteenth- and nineteenth-century roots, it was not until the middle of the twentieth century that the idea of development was popularised to the extent that it inspired an international political agenda and distinct field of study. In the wake of the trauma of the Second World War, a new international order was negotiated that sought to improve international economic collaboration, eliminate conflicts between nations and maintain world peace. A new family of multilateral institutions were established to serve these ends, including the United Nations, the IMF, the World Bank Group (including the International Bank for Reconstruction and Development) and the General Agreement on Tariffs and Trade (GATT), set up in 1947 to promote free trade between nations – a precursor to the present World Trade Organization (WTO). Initially, these institutions focused on facilitating the reconstruction of Europe in concert with the European Recovery Plan (otherwise known as the Marshall Plan), which was developed and financed by the United States and funnelled billions of dollars into post-war reconstruction efforts over the course of four years. However, there was also a push for the decolonisation of Africa, the Middle East, Asia and some remaining European possessions in the Americas. President Harry S. Truman's inaugural address on 20 January 1949 illustrates well the sentiments of the time:

> We must embark on a bold new program for making the benefits of our scientific advances and industrial progress available for the improvement and growth of underdeveloped areas. The old imperialism – exploitation for foreign profit – has no place in our plans. What we envisage is a program of development based on the concepts of democratic fair dealing.
>
> (cited in Cowen and Shenton 1996: 7)

Truman's speech ushered in the first development decade. The fact that European nations (and Japan) had been successfully reindustrialised after the war with international assistance fed optimism that this approach could be extended elsewhere in the world.

Roughly speaking, one can discern four post-war periods in development theory, policy and practice. The 1950s and 1960s were decades of optimism dominated by 'modernisation' theory, which emphasised state-driven industrial expansion to accelerate economic growth and development, modelled, to some extent, on the Marshall Plan. The 1970s were a decade of global economic turmoil attended by wide-ranging critiques of the theories, policies and practices advocated by modernisation theorists and mainstream international development actors. Theories of dependency emerged alongside critiques of the power imbalances between various actors within and between nations. This led to a shift in policy focus from emphasising production and growth to attending to people and poverty. By the beginning of the 1980s, however, a new intellectual agenda known as 'neo-liberalism' gained momentum. Through the 1980s and 1990s there was a push to 'roll-back' states and reduce their disruptive role in markets and people's lives. But by the turn of the millennium it became clear that free-market reign was failing to produce the promised benefits. In the contemporary period, development theory and practice are increasingly focused on institutions and governance, or the incentive structures and relationships that underpin the performance of states and markets. This concern with making states *and* markets work better has been further bolstered by rising security concerns. Following the terrorist attacks of 11 September 2001, poverty, inequality and the global security threats posed by 'failed states' have been linked to the failures of development in low- and middle-income countries. As a result, the development agenda has become increasingly enmeshed with concerns about national security. Table 1.1 provides a schema of this periodisation with a crude summary of the mainstream policy agenda attending each period. We briefly explore each of these periods in turn.

Modernisation theorists, largely based in the United States and the United Kingdom, conceptualised development as a dynamic process of economic and social transformation driven by a class of enlightened political elites. It involved a shift from agricultural to industrial production, from rural to urban habitation, and from 'traditional' to 'rational' socio-political values (Thomas 2000b; Willis 2005). Countries that remained largely agricultural, or that were deemed 'traditional' in their forms of social organisation and values, were referred to by modernisation theorists as 'undeveloped' or 'backward'. Development was understood as a linear process of cumulative change and the key problematic addressed by scholars in the post-Second World War era was how to set the wheels of development in motion – or to use the terminology of American economist and modernisation theorist Walter Rostow (1960), how to achieve 'take-off'. In his book *The Stages of Economic Growth: A Non-Communist Manifesto*, Rostow (1960) articulated a process of development not dissimilar in direction to that proposed by Marx although, as his title suggests, with very different goals in mind. In his view, all societies could advance from a traditional, low-productivity state to one of capitalist high mass-consumption, such as had been achieved by the United States, so long as successive steps were followed to ensure the 'preconditions for take off', then economic 'take off', and finally a 'drive to maturity'.

Table 1.1 ***Key trends in mainstream development theory, policy and practice***

1950s–1960s Modernisation/ developmentalism	*1970s Dependency: empowerment critique*	*1980s–1990s Neo-liberalism*	*2000s Institutions and governance/security and development*
Industrial expansion through investment and interventionist economic policies; poverty reduction through economic growth and residual approaches to welfare	Growth with redistribution; increased emphasis on poverty reduction through supporting production and basic needs; participation and empowerment	Structural adjustment: privatisation, liberalisation and deregulation of markets; reduction of subsidiaries and introduction of cost recovery	Focus on improving accountability, transparency, rule of law, security of property rights, bureaucratic efficiency and state effectiveness; linking of security and development agendas

In order to achieve this, it was argued that late developers would have to mobilise savings towards a process of capitalist industrialisation, which would provide the critical foundations for this process. It was believed that international aid could make up for shortfalls of capital that limited investment in less economically advanced countries. It was also argued that states should play an active role in cultivating industrial development by investing directly in 'infant industries' and indirectly encouraging expansion and diversification through policies to restrict trade in certain sectors of the economy. Investment and industrial policy were accompanied by large-scale infrastructure projects to build roads, housing, dams and energy infrastructure. The World Bank, in particular, promoted and financed such projects through grants and loans. The success of state-directed development is most closely associated with the 'developmental states' of East Asia, which successfully facilitated a dramatic economic transformation in countries such as Japan, Taiwan and South Korea over the course of the latter half of the twentieth century. This East Asian 'economic miracle' has been attributed to effectively 'governed markets' where there was a 'synergistic connection between a public system and a mostly market system' (Wade 2004: 5). In other words, government officials and politicians were able to construct and apply economic rules that advanced technological development and long-term capital growth.

In Latin America, South Asia and Sub-Saharan Africa, there was also a post-war economic boom largely driven by state-directed investments and activist policies but here developmentalism had a more chequered fate. Although there were some notable successes, by the 1970s it had become clear that the fruits of progress did not necessarily 'trickle down' from the wealthy to the poor as modernisation theory had predicted, and that industrial policies were facilitating corruption and graft at the highest levels of government. Poverty continued to rise, urban informal settlements were growing and state-driven industrialisation was failing to generate sufficient jobs to absorb a growing non-agricultural labour force. The limits of developmentalism were thrown into sharp relief in the1970s when an oil price shock sent devastating ripples through the world economy. It was also in the 1970s that a wide range of critiques – many emerging from those parts of the world where developmentalism had failed – were levelled against the international development agenda more generally.

The first and most fundamental critique of modernisation theory came in the late 1960s from dependency theorists, who argued that countries were not 'backward' or 'undeveloped' but that they were deliberately 'underdeveloped' by the international capitalist system, in a process that helped the advanced economies extend and maintain their prosperity at the expense of weaker economies. According to the dependency perspective, low- and middle-income countries had been incorporated into this system on terms beyond their control and against their general interests during the colonial era and post-independence period. So-called 'modernising' elites were dubbed a 'comprador' class that worked to advance their own interests and those of international capital, rather than the general interests of national economies and citizens.

Dependency theory is most closely associated with the writings of the German-American economist, Andre Gunder Frank, who developed his ideas while working with the Economic Commission for Latin America (CEPAL). Through detailed historical studies of Chile and Brazil, and drawing on both Latin American development debates and the writings of North American Marxian Paul Baran (1957), Frank argued in *Capitalism and Underdevelopment in Latin America* (1967) that development in low- and middle-income countries was not possible because economic surplus was extracted from the global periphery (i.e. less industrialised economies) by or on behalf of the metropolitan centres of wealthy countries. Dependency theorists also saw the whole development project as a grand design aimed at masking the underlying extractive intentions of rich and powerful nations and firms (Cardoso 1972). As an analytical framework, dependency theory was very compelling, particularly to policy makers in developing countries who had to answer to domestic constituencies for their failures. However, it was problematic in that it never really offered an alternative path for development; clear goals were not articulated, nor were any concrete policy alternatives (Palma 1981). Without an accompanying programme of action, dependency theory was confined to the arena of intellectual debate and never really established an operational framework.

However, other perspectives began to emerge in the 1970s that changed the way the practice of development was understood and pursued. Macro-level concerns about economic transformation were augmented by more micro-level concerns about the lived experience of the poor. Increased attention was paid to the notion that the ultimate goal of development is improvement in the lives of people, and the expansion of production and exchange in national economies was increasingly viewed as the means rather than the end goal of development. For the most part, agents of 'development' were thought to have lost sight of the consequences of development interventions for different segments of populations and were failing to recognise the needs and potential of poor people. The prevailing notion that development was essentially equal to economic progress was challenged, and poverty began to be conceptualised in more holistic terms related to power and social relationships. Scholar/professionals such as Robert Chambers (1983) argued that development should be turned on its head: development initiatives should be more participatory, more attentive to local voices, and respectful of indigenous knowledge and the expertise of people on the ground. In other words, the practice of development should seek to empower the most vulnerable by giving them greater control over development initiatives. Development researchers and professionals also began to recognise the importance of gender and development, leading to the first UN Conference on Women in Mexico City in 1975. Feminists pointed out the differential impact of development interventions on women and men, arguing that development policies had variously ignored, harmed or used women instrumentally and this had led to outcomes that were biased towards the interests of men (Boserup 1970; Elson 1991; Kabeer 1994; Moser 1993). And environmentalists began to be critical of the development discourse for its relentless promotion of economic growth without paying due heed to ecological consequences (Redclift 1987). These myriad critiques introduced many new concepts and phrases to the development lexicon that remain with us today, such as 'empowerment of the poor', 'participation', 'gender sensitivity' and 'sustainable development'.

By the end of the 1970s, mainstream development thinking (in terms of both policy and practice) had absorbed many of these critiques but ultimately came to be dominated by a new international political-intellectual agenda that was not confined to the concerns of less economically advanced nations. The global economic crisis of the 1970s, which affected rich and poor countries alike, was followed in the United Kingdom and the United States by the elections of Margaret Thatcher and Ronald Reagan, both strong proponents of minimal state intervention in economic affairs. This contributed to a shift in development thinking and policy in the 1980s that was sufficiently drastic for the British economist John Toye to have called it a 'counter-revolution' (Toye 1987). The era of developmentalism gave way to one of free market liberalism, or 'neo-liberalism' – the label commonly used to describe development policy during the last two decades of the twentieth century. Its theoretical genesis predated the 1980s, emerging out of the work of economists such as Milton Friedman and Friedrich von Hayek, who were strong proponents of minimal state involvement in the social and economic affairs of nations. This minimalist-state perspective found its way into development thinking through the work of the California-based economist, Deepak Lal. Much like Adam Smith, he promoted laissez-faire economic policies, arguing that economic growth should be left to the market. He was of the view that in an imperfect world, imperfect market mechanisms were better than imperfect state interventions (Lal 1985). While previously the state had been seen as a key driver of development, it was now seen as the primary obstacle. State interventions, it was argued, distorted the natural functioning of free markets leading to development failures. If the previous decades can be characterised as concerned with 'getting policies right', the new paradigm revolved around the idea of 'getting prices right' by eliminating state-created market distortions. By the end of the 1980s international development policy and practice had been thoroughly revamped based on these neo-liberal ideas.

The concomitant policy response advocated by the international financial institutions (IFIs) was macroeconomic reform and a significant downsizing of the role of the state in cultivating industrialisation. Low- and middle-income countries were encouraged to pursue a wide range of reforms including: privatisation of state enterprises, fiscal policy discipline (e.g. through reducing public sector employment), liberalisation of international trade (e.g. reducing tariffs on imported goods), liberalisation of financial markets to facilitate foreign direct investment, tax reform, strengthening of private property rights and deregulation of domestic markets. These reforms, designed to create more competitive, internationally integrated markets that would attract private sector investment, came to be collectively known as the 'Washington Consensus'. This ambitious reform agenda was pushed on to low- and middle-income countries through the World Bank and IMF's Structural Adjustment Programmes (SAPs). These included 'conditionalities' under which countries had to agree to undertake some combination of these reforms in order to qualify for desperately needed grants and soft loans offered through SAPs. The Washington Consensus recipe was applied equally to countries across Africa, Asia and Latin America as well as to those of Eastern Europe and the former Soviet Union.

The stranglehold of neo-liberal market orthodoxy seems, at the present moment, to be loosening. Just as developmentalism (and developmental states) demonstrated a patchy record, so too have market-oriented reforms. Many have referred to the 1980s as the 'lost decade' of development, and progress in reducing poverty in the 1990s has been markedly uneven across countries and regions. This has inspired a new research and policy agenda that revolves around institutions and governance – the underlying incentive structures and relationships that determine the performance of states and markets alike. It is now recognised that the capacity of governments to provide public goods, effectively regulate markets, formulate sound policies and be responsive to citizens' needs varies markedly across countries. It is also recognised that markets do not function efficiently without effective state regulation. In the contemporary literature, the preoccupation is with 'getting institutions rights', with institutions defined as the 'rules of the games' that structure social interaction (North 1990). These can be formal (i.e. laws) or informal (i.e. social norms of behaviour), and they can affect the investment decisions of individuals and businesses, the efficiency of market transactions and the extent to which governments are accountable to their citizens. Governance is a more general term that refers to the relationship between citizens and governments at different levels; to the ways in which decisions that affect society are made and implemented. In practice, this new perspective translates into policies and interventions that seek to improve the performance of states and markets alike by identifying and promoting the kinds of incentive structures that encourage efficiency and accountability. There is also a general consensus among scholars from many disciplines that the colonial era had profound and long-lasting impacts on the quality of institutions and governance in most countries that now fall into the category of low- and middle-income (Acemoglu *et al.* 2001; Mamdani 1996). In other words, the way in which countries came into being, and the way in which particular communities and societies were integrated into the global economy, has had long-term consequences for their domestic development.

It must be noted that these general trends in mainstream development theory, policy and practice cannot really be neatly periodised. The influence of modernisation theory in development studies (and particularly development economics) remains strong, albeit in modified form. Contemporary accounts of the varying successes and failures of developmental states, such as Amsden (2001), Evans (1995), Kohli (2004) and Wade (2004), in many ways reflect the interest of modernisation theorists with the catalytic potential of states without the baggage of assuming that development progresses through definable stages. Others offer narratives of progress and differentiation that reflect classical modernisation thinking. For instance, historian David Landes (1998), recalling the work of early-twentieth-century sociologist Max Weber, claims that 'if we have learned anything from the history of economic development, it is that culture makes all the difference' (Landes 1998: 516). In doing so, he recalls modernisation theorists' concerns with the transition from 'traditional' to 'modern' values.

We can also discern echoes of dependency theory in contemporary debates about globalisation. Broadly speaking, globalisation refers to processes of economic integration that have opened up more economies and societies to world markets. While some argue that globalisation offers unprecedented opportunities for low- and middle-income countries, others, who might be viewed as successors of the dependency theorists, point to the fact that globalisation serves to entrench existing imbalances of power and opportunity within and between nations, particularly when countries cannot dictate the terms of their engagement with the global political economy (Stiglitz 2002).

Finally, concerns about empowerment, participation, gender and sustainability raised in the 1970s have, to some degree, become part of mainstream development discourse and practice. The importance of participation in the design and implementation of development initiatives is now widely accepted and justified from both normative and instrumental perspectives. From a normative perspective, local ownership and participation are seen as intrinsically good, empowering people to find their own solutions to their problems and hence enhancing their real freedoms. From an instrumental perspective, it is argued that participation and local ownership produce better outcomes by harnessing the knowledge and energies of local populations while enhancing their commitment to development projects. For example, the World Bank has moved away from the use of conditionalities and now encourages countries seeking development assistance to produce Poverty Reduction Strategy Papers (PRSPs) that outline the policies and actions that a government intends to adopt in order to reduce poverty and cultivate development. PRSPs are ostensibly the product of a broad consultation exercise that involves government actors and representatives from civil society and the private sector. Whether or not PRSPs do in fact represent a participatory approach to policy making in practice is subject to debate, but the approach does reflect the broader trend towards emphasising local ownership and participation. Concerns about gender have been mainstreamed into development policy and practice at many levels (albeit with varying degrees of success), and the long-term environmental consequences of development initiatives are increasingly taken into account. Indeed, sustainability is now firmly entrenched as one of the major concerns of the twenty-first century, along with security, poverty eradication and growth.

There are also those who continue to question the practices and motivations of those engaged in development. Anthropologists, in particular, have been critical of development interventions and their consequences. For example, in his study of development assistance in Lesotho, James Ferguson (1990) characterised the development industry as an 'anti-politics machine' that saps local energy and insight through the serial failures of external managers who promote technical interventions that demonstrate a startling ignorance of the historical and political realities of the places they intended to help. In *Fear of Small Numbers*, Arjun Appadurai (2006) portrayed development as a vehicle used by certain governments to control social and political dissonance. Appadurai advocates development from below, or what he calls 'grassroots globalization', being the joined-up aims and activities of local organisations and transnational advocacy movements. More radical critics have questioned the salience of the development project altogether. Influenced by dependency theory, the work of anarchist Ivan Illich (1992), who characterised development as 'planned poverty', as well as post-structuralist philosophers such as Michel Foucault (1988), who sought to understand the subliminal dimensions of power, for example through language and discourse, 'post-development' thinkers employ discourse analysis to destabilise development ideas and practice. They interrogate the ways in which development discourse has served the interests of advanced economies at the expense of the majority world. Colombian-born anthropologist Arturo Escobar has described developmentalism as a bogus discourse with illusory goals in which material achievement continues to be advocated in a world where poverty and inequality remained disturbingly widespread (Escobar 1995).

These critics make an important contribution to development theory and practice by debunking the many myths of development, but they suffer from the same problem as dependency theory in that they do not provide policy alternatives. Given the poor track record of development since the 1950s, some fresh thinking is definitely in order. Yet, as Alan Thomas has pointed out, throwing out the term and abandoning the project will not alleviate the very real problems of 'poverty and powerlessness, environmental degradation and social disorder' (Thomas 2000a: 21) that continue to affect billions of people across the world. While we recognise the contested nature of development in theory and in practice, we are of the view that development is both desirable and possible. And, following Sen (1999), we take the position that development in practice is ultimately about trying to contribute to the expansion of real freedoms to more people across the world.

The legacy of 'urban bias' in development studies

Where have cities been situated in the evolving theories, policies and practices of development? Over the years, cities have variously been seen as by-products of economic change, drivers of modernisation and as obstacles to development. In the early development decades, urbanisation was understood by modernisation theorists to be a natural consequence of economic growth, and investments in urban infrastructure and industry were encouraged. The work of economist Arthur Lewis was particularly influential at the time, informing development policy and providing a foundation for much of the later theoretical work of development economics. Lewis (1954) proposed a dual-sector model of economic development in which low-productivity labour in the 'traditional' (rural) agricultural sector is transferred to the 'modern' (urban) industrial sector as an economy grows. Furthermore, Lewis proposed that the supply of cheap labour from the traditional sector was essentially unlimited in less advanced economies and that the expansion of the modern sector would encourage migration from the former to the latter. This migration would keep wages down, allowing for the high profits necessary for further capitalist expansion.

The Lewis model was complemented by the work of economist Albert Hirschman. In *The Strategy of Economic Development* (1958), Hirschman argued that the expansion of the modern sector was driven by key economic sectors that necessarily involved spatial concentration and inequality. Uneven development and polarisation were, in his view, inevitable features of countries in the early stages of economic growth but would eventually be overcome by the 'trickle down' of benefits to surrounding areas. Gunnar Myrdal (1957) was less optimistic, arguing that spatial differentials and inequalities, once established, would be difficult to overcome.

Regional planning theorists and professionals at the time thought that spatial inequalities could be mitigated through targeted interventions. For example, the North American planner John Friedmann (a pioneer in the field of regional planning) devised policies to encourage the growth of medium-sized cities in peripheral regions and the establishment of urban 'growth poles' (Friedmann and Alonso 1975). It was generally believed that urban centres could be used to drive regional development in less economically advanced nations, and that over time cities and urban systems would naturally evolve in such a way as to encourage the integration of national economies. However, beginning in the 1950s, concerns emerged about the pace and scale of urbanisation in Latin America, Africa and Asia and the term 'over-urbanisation' was coined.

Over-urbanisation refers to the relationship between a nation's level of urbanisation and the distribution of its labour force across sectors (i.e. agricultural, manufacturing, industry and services). A country is said to be over-urbanised (or under-urbanised) if the ratio between its level of urbanisation and the percentage of its labour force employed in industry deviates significantly from the observed ratio in advanced economies. At a UN conference in 1956 it was claimed that in over-urbanised countries, 'urban misery and rural poverty exist side by side with the result that the city can hardly be called "dynamic," as social historians of developed countries generally described the process of urbanization' (cited in Sovani 1964: 113). Cities across Africa, Asia and Latin America were increasingly characterised by burgeoning squatter settlements, shanty-towns and *favelas*, fuelling fears about the negative social and political impact of urban growth.

These concerns were augmented by dependency theorists, who were critical of the notion that development inevitability entails inequality, unbalanced growth and the 'natural' evolution of urban systems over time. They argued that towns and cities in less economically advanced nations did not diffuse development out to their hinterlands but towards Western economies (Potter 1992). In effect, towns and cities were seen as hubs in an international system of extraction where primary goods would flow from peasant producers in the countryside through market towns, regional centres and national capitals to international metropoles, creating a pattern of dependent, parasitic urbanisation.

By the end of the 1970s, far from seeing rural–urban migration as the economic bonus posited by Lewis, a growing number of scholars and policy makers concluded that rapid urbanisation was a problem. Cities were no longer seen as drivers of regional and national economic modernisation, but rather parasitic, dystopian spaces. This growing negativity about urbanisation and cities crystallised in the form of the 'urban bias thesis' associated with the works of Michael Lipton and Robert Bates. In the dramatic opening paragraph of his book *Why Poor People Stay Poor: Urban Bias in World Development* (1977), Lipton drew a line in the sand between rural and urban development and claimed that cities were impeding development:

> The most important class conflict in the poor countries of the world today is not between labour and capital. Nor is it between foreign and national interests. It is between the rural classes and the urban classes. The rural sector contains most of the poverty, and most of the low-cost sources of potential advance; but the urban sector contains most of the articulateness, organization and power. So the urban classes have been able to 'win' most of the rounds of the struggle with the countryside; but in so doing they have made the development process needlessly slow and unfair.
>
> (Lipton 1977: 1)

Lipton's thesis captured the imagination of development scholars and practitioners at the time and for many years to come. He argued that governments in developing countries imposed price distortions that favoured urban over rural areas. According to Lipton, prices were 'twisted' against rural dwellers in two ways. First, countries overvalued their currency in order to lower the cost of foreign imports. This largely benefited urban elites who depended on Western-produced machinery or inputs for their industries and who desired imported consumer goods. However, it had the concomitant effect of lowering the prices that farmers were able to command for their agricultural exports. A second way in which prices were 'twisted' against farmers was by governments commonly using their power to buy up a country's agricultural products, for example through agricultural marketing boards. This provided urban dwellers with cheap food supplies but hit farmers in the pocket. In acting as a single or dominant purchaser, governments prevented competitive markets and pricing from arising that would have benefited rural agricultural producers (see Box 1.2).

Robert Bates (1981, 1988) extended Lipton's thesis in the 1980s in his analysis of agricultural systems in Sub-Saharan Africa. Here he posited that in addition to an economic playing field that was far from level, the political power of cities thoroughly eclipsed the voice of smallholding agriculturalists. Fear of urban food riots and of upsetting proximate and organised urban groups meant that governments acted not as vehicles for maximising the social welfare of all their citizens but rather as agents that 'accommodate the demands of organised private interests' (Bates 1988:

Box 1.2

The 'urban bias thesis' and critiques

The urban bias thesis emerged in the late 1960s and was advanced most forcefully by economist Michael Lipton. Simply stated, urban bias manifests in public policies that favour the interests and welfare of urban populations. Lipton claimed that such a bias is inefficient as well as unfair. In particular, agricultural pricing policies and public investments are skewed to ensure cheap food and better services for urban populations, thereby constraining economic development and welfare improvements in rural areas. This bias is said to have arisen in the 1950s and 1960s due to mainstream development theory and policy, which advocated urban-based industrial expansion, and perpetuated by the political power of 'urban classes'.

The urban bias thesis has been critiqued on several counts. First, rural and urban settlements exist on a continuum and cannot be categorically opposed, particularly given the regular flows of people between settlements and regions. Second, the linkages and complementarities between rural and urban economies suggest that public policy making is not a zero-sum game. Third, intra-urban inequality betrays the notion that urban residents constitute a distinct 'class' that works corporately to advance its own interests at the expense of a rural 'class'. With urban poverty rising rapidly in low-income countries, it is increasingly apparent that 'urban bias' is an analytically problematic concept.

121). Both Lipton and Bates were of the view that rural areas received too little by way of social sector spending, for example on health and education, relative to the size of their populations and their need.

A growing sense that cities were growing too fast, that they were failing to spread the benefits of growth, and that urban populations were unjustly favoured over rural ones contributed to a significant change of focus for researchers and policy makers alike. Rural development was placed at the very centre of development discourse and practice. For those concerned with economic growth, it was argued that agricultural development must precede industrial development. For those concerned with poverty reduction it was pointed out that most people in the world, and the overwhelming majority of the poor people in the world, lived in rural areas. There was a shift away from industrialisation strategies and towards integrated rural development policies. The 'green revolution' initiated in the late 1960s, which sought to increase agricultural productivity in low- and middle-income countries by introducing more productive crop varieties and better cultivation practices, was wholeheartedly embraced.

These debates and policies were complemented by strategies to manage rural–urban migration in order to prevent over-urbanisation. It was argued that the best way to control urbanisation was to raise the standard of living in rural areas to the equivalent of that enjoyed by the residents of cities. It was thought that if agriculture could be rendered more productive and if adequate social services and amenities could be brought to rural areas, people would not want to migrate to cities. Michael Todaro's model of migration (Harris and Todaro 1970; Todaro 2000), which sought to explain why people continued to migrate to urban areas despite rising urban unemployment, was particularly influential. He argued that rural–urban migration was rational because it was based on anticipated rather than actual benefits and that it was inevitable because of the imbalance in economic opportunities between rural and urban areas in most low- and middle-income countries. Todaro's model reinforced the policy implications of Lipton's urban bias thesis, for example eliminating price distortions between town and countryside, expanding small-scale, labour intensive industries and, significantly, reducing the imbalances in urban–rural employment opportunities and wages:

> It is vitally important that imbalances between economic opportunities in rural and urban sectors be minimised. Permitting urban wage rates to rise at a greater pace than average rural incomes will stimulate further rural–urban migration in spite of rising levels of urban unemployment. This heavy influx of people into urban areas not only gives rise to socio-economic problems in the cities but may also eventually create problems of labour shortages in rural areas, especially during the busy seasons. These social costs may exceed the private benefits of migration.
>
> (Todaro 2000: 310)

These ideas struck a chord with many national leaders in low- and middle-income countries at the time, including Mao Ze Dong in China and Julius Nyerere in Tanzania, both of whom sought to address the problems faced by impoverished peasantries by reducing differentials in rural–urban welfare. Legal restrictions on migration, forced relocation, slum clearance and rural development stratgies were all employed to stem the tide of rural–urban migration. Some of the more extreme examples include Indonesia's transmigration programme, which over the last quarter of the twentieth century resettled millions of people from areas of high to low density, or China's *Hukou* system of household registration that from 1955 onwards required people to obtain permission to move from their places of birth and residence (Lall *et al*. 2006). Most pernicious perhaps was the monitoring of people's movements in South

Africa under the Apartheid Regime (1948–1994). In an effort to ensure that black South Africans remained primarily in the rural reserves, they had to carry passes and were considered as temporary sojourners in cities, welcome only as workers where they were tolerated for their service to white-owned enterprises and households.

The urban bias thesis (UBT) and the dominance of the rural development paradigm in development studies, policy and practice has led to a general neglect of cities since the late 1970s. This neglect is apparent, for example, in contemporary Poverty Reduction Strategy Papers. Two reviews of PRSPs reveal that nearly all have a strong emphasis on the relative importance of rural poverty and development while neglecting urban poverty altogether or demonstrating a generally poor understanding of urban poverty and development issues (ComHabitat 2005; Mitlin 2004a).

Given current demographic trends (discussed below), the inevitability of urbanisation in the course of economic development, the undeniable importance of cities to economic growth (see Chapter 3) and the evident increase in urban poverty and inequality (see Chapter 4), the continued resonance of the urban bias thesis is difficult to understand. Corbridge and Jones (2006) have argued that its durability can be explained in part by the merits of the argument at the time, but point as well to the limits of its continued salience as an analytical or an operational framework for development:

> To sum up: there is much to admire in the urban bias thesis, but the most important parts of it can be restated (as a sort of level playing field argument, for example, or as an argument against some forms of predation) without any resort to a generalised model of city–countryside exploitation. It is misleading to speak of a single urban class exploiting a single rural class . . . To the extent that the UBT has encouraged a neglect of urban poverty and the economic dynamism of many cities in the developing world, it has also had an unwelcome effect on policy. What is required now is a re-balancing.
>
> (Corbridge and Jones 2006: 38)

This book represents our attempt to contribute to this rebalancing by examining both the potential for urban development to contribute to national development and some of the key challenges confronting development scholars and professionals concerned with urban poverty. Given current trends, the need for this rebalancing is more pressing than ever.

The first urban century

In 1900, just 13 per cent of the world's population lived in urban areas. By 1950 that number had risen to 29 per cent, and in 2005 some 49 per cent of the world's population was urban. The United Nations projects that 4.9 billion people will live in cities by 2030, representing 60 per cent of the global population (United Nations 2008) (Figure 1.1).

This trend towards an increasingly urban world is being driven primarily by urbanisation and urban growth in low- and middle-income countries – particularly in Africa and Asia. Urbanisation refers to a rising proportion of a nation's population living in urban areas; urban growth refers to an increase in the absolute size of a nation's urban population. The distinction between these terms is important to keep in mind. It is possible, for example, for a nation to urbanise with no urban growth if the absolute rural population declines; it is also possible to have rapid urban growth (a rapid increase in the absolute size of a nation's urban population) without urbanisation if the absolute size of the rural population grows as fast or faster.

Table 1.2 summarises the trends in urban growth and urbanisation by region. Nowadays, Africa and Asia exhibit the highest rates of urban

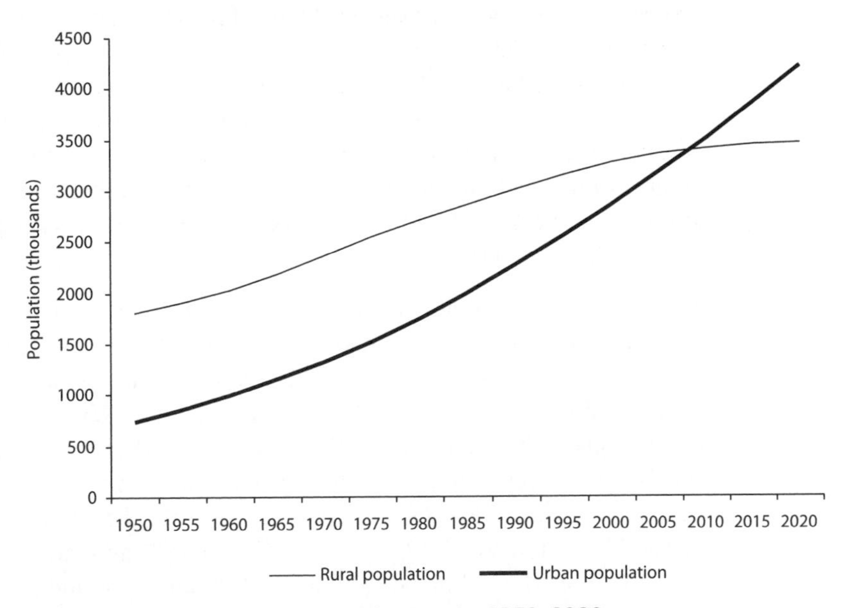

Figure 1.1 Rural and urban population trends, 1950–2020

Source: United Nations (2008)

Table 1.2 ***Urbanisation and urban growth rates by region***

	Average annual rate of urban growth			*Average annual rate of urbanisation*		
	1950–1970	*1970–1990*	*1990–2010*	*1950–1970*	*1970–1990*	*1990–2010*
Africa	4.86	4.32	3.52	2.44	1.52	1.10
Asia	3.59	3.70	2.78	1.50	1.70	1.43
Europe	1.92	1.05	0.21	1.02	0.58	0.15
Latin America	4.30	3.25	2.03	1.60	1.07	0.59
Northern America	2.23	1.12	1.45	0.72	0.11	0.43

Source: United Nations (2008)

growth and urbanisation, and they will continue to do so well into the twenty-first century. Latin American and Caribbean countries experienced similarly high rates of urban growth and urbanisation in the middle of the twentieth century. This region is now fully urbanised, and hence urbanisation rates have slowed. However, many countries (and cities) in the region continue to experience relatively high urban growth rates. High rates of urban growth present a significant challenge to national governments seeking to effectively manage rapidly expanding cities.

From the perspective of urban development policy and practice, the most alarming fact is that *virtually all of the world's population growth over the next few decades will be absorbed by cities in low- and middle-income countries*. While rural populations will continue to rise over the next decade, in 2019 that trend is expected to reverse so that the absolute number of people living in rural areas will begin to decline while the absolute number of people living in cities will continue to rise rapidly (United Nations 2008).

Two consequences of these trends are beginning to catch headlines: the proliferation of mega-cities and the burgeoning 'slum' settlements across low- and middle-income countries. Mega-cities are defined as urban agglomerations of 10 million inhabitants or more. In 1950 there were just two mega-cities in the world (New York–Newark metropolitan area and Tokyo); in 2005 there were 20 mega-cities in the world, 15 of which were in low- and middle-income countries; by 2015 there will be 22 mega-cities in the world, 17 of which will be in low- and middle-income countries, including Mumbai and Mexico City (at 22 million inhabitants each) and São Paulo (at 21 million). Yet these cities are not the fastest growing, nor are they home to a disproportionate number of people. In fact, 51 per cent of the world's urban population live in cities of just 500,000 people or less; 90 per cent live in cities of 10 million or less. Small- and middle-size urban settlements will continue to absorb the bulk of urban growth in the coming decade (United Nations 2006).

Rapid urban growth is being accompanied by the growth of sprawling settlements characterised by poor infrastructure and housing – or 'slums' in the current discourse. Until recently, the term 'slums' was used to refer to decaying, overcrowded and densely occupied areas, often in city centres. They were seen as distinct from informal settlements or shanty-towns that are more often found on the periphery of urban centres. However, the term 'slum' is now commonly used to embrace all areas characterised by poor quality dwellings and inadequate infrastructure, whether downtown tenements or irregular peripheral settlements. Slum settlements are defined by lack of secure tenure, poor (or non-existent) basic infrastructure such as water and sewerage, and poor-quality dwellings. UN-Habitat (2007) estimates show that approximately 1 billion people live in slums nowadays. They are vulnerable to disease, violence, environmental hazards and social, political and economic exclusion – and of course this number is predicted to rise in the years ahead. Figures 1.2 and 1.3 give an idea of regional trends. The largest slum populations in

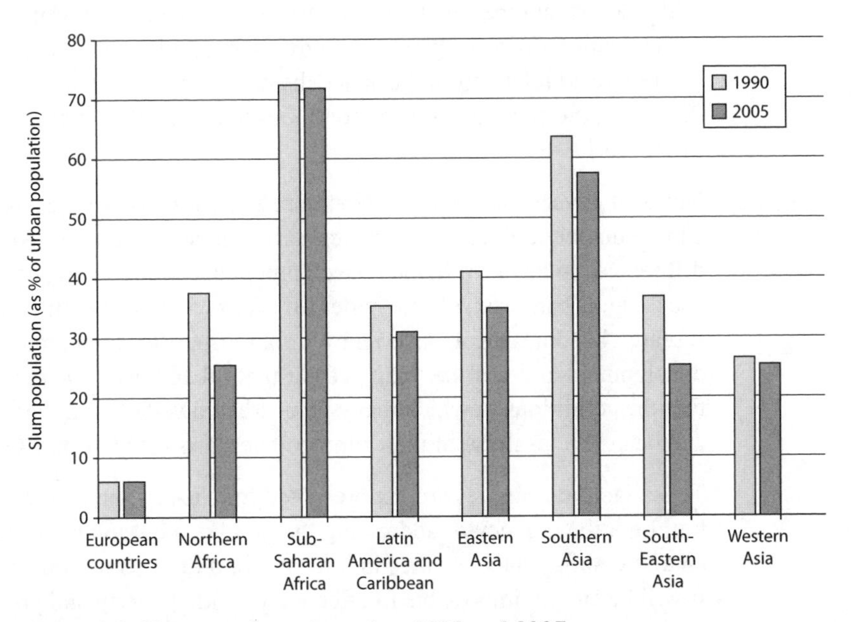

Figure 1.2 Slum prevalence by region, 1990 and 2005

Source: UN-Habitat (2007)

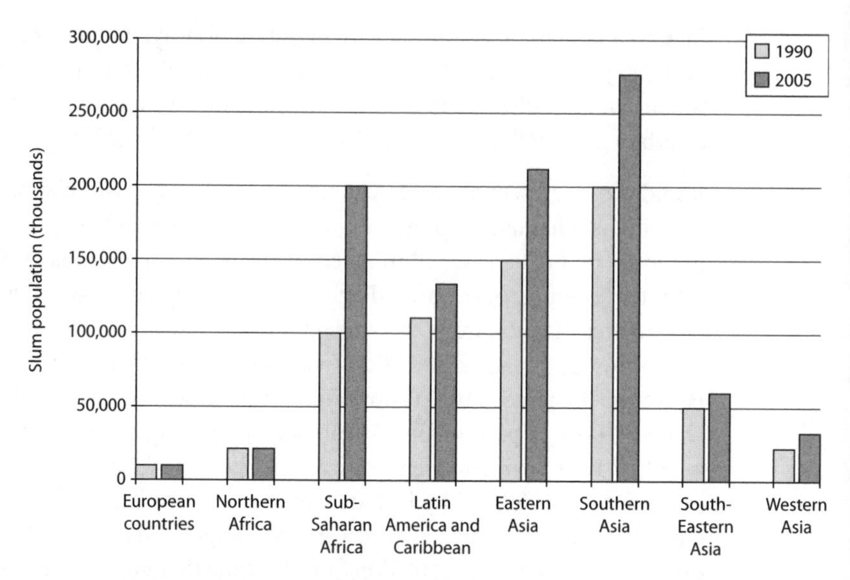

Figure 1.3 Slum population by region, 1990 and 2005

Source: UN-Habitat (2007)

the world in relative terms are in Sub-Saharan Africa, but Asia is home to the largest number of slum dwellers by far (in absolute terms). And while the proportion of global slum dwellers relative to urban populations has remained relatively stable since the early 1990s (and in some cases declined), the absolute number continues to rise in almost every region in the world.

National governments and development agencies have in the past used a simple distinction between rural and urban areas to monitor geographical differences in levels of human development and welfare. Generally speaking, urban residents are better off. Yet as slums grow and more research becomes available on intra-urban disparities in human development and well-being, it is clear that urban poverty is rising and that the conditions in which many poor urban dwellers find themselves can be as dire as those of their rural counterparts (see Chapter 4).

These facts and figures are not presented for shock value, but rather to highlight the urgency of addressing the rapid demographic and social changes taking place in cities in low- and middle-income countries. It will be simply impossible to effectively tackle poverty and promote sustainable development in the first urban century without an understanding the unique characteristics of urban poverty and social disadvantage, as well as the ways in which rapid urbanisation and urban growth can be harnessed as positive forces for development.

Cities and development

Having defined our terms and provided a brief history of development theory and practice, we now turn to an in-depth exploration of the cities-and-development nexus. The following chapters are organised around key themes that stand at the intersection between urban studies and development studies: urban history, urbanism and economic development, urban poverty, managing urban environments, human security in cities, and finally the processes of urban planning, governance and politics that largely define the prospects for building inclusive, productive and sustainable cities.

Chapter 2 provides a sweeping overview of urban history beginning some six thousand years ago when the first large, permanent human settlements appeared in Mesopotamia. Tracing the evolution of urbanism through the twentieth century highlights the deep historical origins of the relationships between the establishment of cities, the process of urbanisation, economic development and the transformation of human societies. We examine not only the relationship between the demographic process of urbanisation and industrial transformation in Europe, but also how rising European powers set in motion very different patterns of urban development in other regions of the world through imperialism and colonisation. In the last section of Chapter 2 we look at how contemporary demographic processes, combined with historical legacies of distorted political and economic integration into the global political economy, are generating patterns of urbanism that present significant challenges for planners, policy makers, and others concerned with combating urban poverty, generating wealth and promoting inclusive and sustainable cities.

Chapter 3 turns to one of the most persistent themes in development studies: economic development. Here we examine the potential of urban settlements to be uniquely dynamic economic spaces. The size, density and heterogeneity of urban populations create conditions conducive to economic growth and development. But cities are never islands of economic activity; they are always dependent on flows of people, goods, money and ideas from their surrounding regions and other cities, near and far. Indeed, cities serve as critical nodes in regional economies and the global economy more generally. Understanding the economic flows between rural and urban areas, between cities, and between city-regions across the globe is an important preoccupation of urban scholars. For development scholars, particularly those concerned with rising urban poverty, the 'informalisation' of urban economic activity – including flows within and between cities – is a key area of research. Does the spread of unregulated economic activity within and between cities represent a positive response on the part of the urban poor to difficult circumstances, or the failure of governments to generate growth and employment?

In Chapter 4 we note that reliance on informal economic activity is one of the many characteristics of urban poverty and vulnerability. Poverty can be simply defined as a lack of something, but it is better understood as a state of persistent vulnerability, which in an urban context results from reliance on a monetised economy, poor job prospects in unregulated markets, poor infrastructure, poor health, insecurity of tenure, social fragmentation and endemic violence. These conspire to create a cycle of impoverishment that is difficult for individuals to escape and policy makers to break. Interventions generally focus on one of two areas: working to improve the prospects of the urban poor to establish sustainable livelihoods and working to improve the quality of housing. With few exceptions, targeted interventions aiming to improve the lot of the urban poor have failed to generate sustainable change on a sufficient scale.

Another angle of intervention has been on the critical urban services that determine the quality of life in urban environments more generally. As we show in Chapter 5, efforts to improve water, sanitation and solid waste disposal services have not kept pace with increasing demand due to rapid urban growth. The failure to provide these services efficiently and equitably contributes to the poverty trap experienced by the urban poor, and more generally highlights the challenges of public goods provision in cities. While the negative environmental impacts of poor service provision are generally local, poor public transport systems have both local and global implications. At the local level, traffic accidents, pollution and congestion represent significant costs, placing a drag on the economic potential of urban settlements through their effects on the health of urban residents and the efficiency of economic flows in an urban space. At the global scale, inefficient urban transport systems produce emissions that contribute to global warming. Conversely, climate change is expected to have particularly disastrous effects on cities in low- and middle-income countries, many of which are located in low-lying coastal areas and are vulnerable to rising sea levels and extreme weather events.

We continue our exploration of urban vulnerability in Chapter 6, which addresses the challenges of establishing and maintaining human security in cities. Human security is an absolute prerequisite for development as we have defined it, and urban spaces are uniquely prone to violence and insecurity. The characteristics of size, density and diversity may produce economic dynamism, but they also make cities volatile social and political spaces. When the potential for violence is realised, it can often become endemic. At the extreme, cities can be caught in the crossfire of broader regional or international conflicts, exposing residents to the horrors of urban warfare or the trauma of terrorism. Cities are of both strategic and symbolic significance in regional and global conflicts, making them appealing targets for aggressors. More positively, in the aftermath of conflict cities can serve as strategic and symbolic spaces for reconstruction and reconciliation. Too often this potential is not realised in post-conflict reconstruction efforts.

In Chapter 7 we turn our attention to the processes and relationships that shape urban futures. Unregulated urban growth results in inefficient, inequitable and unsustainable urban environments. Urban planning is therefore of paramount importance. But planning implies a normative vision for urban development. Who shapes this vision and how is it pursued? Given the many actors and forces that converge in cities, planning must be understood as only one piece of an effective approach to urban 'governance', a term that is meant to capture this complexity. In recent years there has been a push for 'good governance' at all levels. One predominant policy strategies to improve governance worldwide has been decentralisation, which is intended to strengthen communication and accountability at local level, but has had mixed effects. It is also a strategy that exists in tension with the demands of large urban centres that require governance structures at a metropolitan scale. Ultimately, however, city futures are shaped by political contestation and leadership, which determine the possibilities for defining and realising a shared vision for inclusive and sustainable urban development.

Finally a disclaimer. This is not a handbook for policy makers and practitioners; nor is it a comprehensive, in-depth treatment of all the issues covered. Instead, we have chosen to sacrifice depth in the interest of breadth, bringing together the wide-ranging fields of urban studies and development studies in order to highlight the need for this vast constellation of interconnected issues and debates to be in conversation with each other. Our hope is that the analysis we provide will serve to sharpen the understanding and engagement of those who have a scholarly or professional interest in cities and development.

Summary

- **Cities are human settlements characterised by their size, density and heterogeneity, and by the social, political, economic and cultural effects of these qualities (or 'urbanism').**
- **Development refers to the way societies change and evolve, and why this differs between countries. The study of development is therefore concerned with analysing processes of *progress* and *differentiation.***
- **Development theory has its roots in the Enlightenment, but the term became popular only after 1945. It is possible to identify four periods that development theory and practice have passed through in the post-war era.**
- **These can broadly be characterised as modernisation/developmentalism in the 1950s and 1960s; dependency theory and empowerment critiques in the 1970s; neo-liberalism in the 1980s and 1990s; and a focus on institutions and governance, as well as the linking of development and security concerns, since the turn of the millennium.**
- **Cities in many low- and middle-income countries grew rapidly in the mid-twentieth century, and the term 'over-urbanisation' was coined. Negative discourses about urbanisation crystallised in the urban bias thesis of Lipton and Bates.**
- **The urban bias thesis placed rural development at the heart of development discourse as policy makers attempted to stem further urban in-migration. This has led to a general neglect of cities while failing to prevent urbanisation.**
- **The world is now more urban than rural and by 2030 60 per cent of the world's population will live in cities. Virtually all population growth in the next few decades will be absorbed by cities in low- and middle-income countries.**

Discussion questions

1. What are the implications of the term 'development' for the study of how societies change over time?
2. Discuss the relationship between the changing nature of mainstream development theory and practice since 1945 and the processes of rapid urbanisation that have occurred in the same period.

3 How might cities be considered important in relation to:

   (a) modernisation theory
   (b) dependency theory?

4 Discuss the impact of the 'urban bias thesis' on cities in low- and middle-income countries.
5 Why have so many efforts to stem the tide of urbanisation proved unsuccessful?

Further reading

Allen, Tim and Alan Thomas (eds) *Poverty and Development in the 21st Century*, Oxford: Oxford University Press.

Bates, Robert (1988) *Toward a Political Economy of Development*, Berkeley, CA: University of California Press.

Drakakis-Smith, David (2000) *The Third World City*, 2nd edn, London: Routledge.

Lipton, Michael (1977) *Why Poor People Stay Poor: Urban Bias in World Development*, London: Maurice Temple Smith.

Sen, Amartya (1999) *Development as Freedom*, New York: Anchor.

Wirth, Louis (1938) 'Urbanism as a Way of Life', *American Journal of Sociology*, 44(1): 1–24.

Useful websites

Cities Alliance, a global coalition of cities and their development partners committed to scaling up successful approaches to poverty reduction: www.citiesalliance.org

Eldis, a development gateway site aiming to share the best in development policy, practice and research: www.eldis.org

United Nations Development Programme: www.undp.org/

United Nations Human Settlements Programme: www.unhabitat.org

World Bank urban development site: www.worldbank.org/urban

2 Urbanisation and development in historical perspective

Introduction

Development – as a historical process that results in the expansion of human freedoms – long predates the rise of development theory and practice. Progressive changes in the social and political institutions that bind human populations together, as well as steady improvements in the material conditions of human societies, began some ten thousand years ago with the rise of agriculture. Shortly thereafter, cities arose, and with them the foundations for civilisations and eventually nation-states. It was not, however, until the nineteenth century that urbanisation took off. Demographic and economic forces created the necessary stimulus that ultimately led to the urbanisation of Europe and other rapidly industrialising nations. Many of these nations used their expanding economic and military power to colonise territories in Africa, Asia and Latin America, a process which had important effects on trajectories of urban development in these areas. By briefly exploring processes of urban change over the *longue durée* we can better understand the patterns of urban development that we find today.

We begin this chapter with an examination of the origins of cities and urbanism. Although there is a general consensus that cities first emerged some six thousand years ago in Sumer in Mesopotamia, this is not an uncontested narrative. Some authors have argued that the first cities appeared earlier. The debate, which we explore briefly below, hinges on our definition of urbanism and is useful in highlighting two essential aspects of urbanism that are easily overlooked in today's increasingly urban and globalised world. Next we survey early urban centres across the world and highlight their functions as political-administrative centres and commercial hubs. In medieval Europe, we find a dynamic tension between urban-based merchant classes and feudal princes that culminated in the formation of nation-states, which came to dominate all other forms of political organisation thereafter. Although the process of state formation in most low- and middle-income countries differed significantly from the European experience, this historical analysis of the process in Europe highlights the importance of cities in processes of political consolidation and change. We then turn our attention to the demographic transition and Industrial Revolution, which marked the beginning of the inexorable process of global urbanisation. While industrialisation transformed human settlements in Europe, colonialism and imperialism radically restructured human settlements in European colonies. Cities were both instruments of colonialism and imprinted by the colonial experience, which left a legacy of urban form and infrastructure designed to segregate people and extract resources. We close the chapter with a summary of contemporary urban transitions (particularly in Asia and Africa), highlighting the changing shape of urbanism in low- and middle-income countries. It is a necessarily non-comprehensive historical survey of urbanisation and urbanism – indeed it barely scratches the surface of the wealth of literature available on the subject. It does, however, draw attention to themes directly relevant to our exploration of the link between cities and development.

The Urban Revolution and the origins of urbanism

For millions of years human beings lived in small bands of hunter-gatherers. Dependence upon nature's produce limited the size of these bands and compelled them to move periodically when local supplies of food were exhausted. About ten thousand years ago the situation changed dramatically. What archaeologists refer to as the Neolithic Revolution (circa 8500 BCE) marked a shift away from human dependence on hunting and gathering and towards livelihood strategies characterised by domesticated agriculture and animal husbandry. This momentous transition was a necessary precondition for the birth of cities, civilisations and ultimately nation-states.

The standard 'origin of cities' narrative suggests that the Neolithic Revolution predated the rise of cities by approximately four thousand years. In this time, farmers made gradual improvements in cultivation, eventually resulting in surplus agricultural production. In particular the cultivation of hard grains that could be produced en masse and stored for significant periods of time reduced the risk of starvation and made it possible to support larger populations. Over time, population growth increased the density of agricultural villages, and rising productivity made it possible not only to hedge against the risk of famine, but also to support a class of individuals who were not engaged in agricultural production. The result was an Urban Revolution, to use Gordon Childe's famous term (1950).

Archaeologists place the birth of cities around six thousand years ago in the Sumer region of Mesopotamia (present-day Iraq) on the plains that lie between the Tigris and Euphrates rivers, where fertile soils and access to waterways for irrigation and transport facilitated surplus agricultural production (see Figure 2.1). However, cities also arose independently at later dates in Africa (1000 BCE), India (1000–400 BCE), China (700–400 BCE), as well as Mexico and Peru (100 BCE). In each case, there is evidence of domesticated agriculture predating the rise of cities, usually by 2000–3000 years (Bairoch 1988).

In all early cities, archaeologists find evidence of a socio-economic order characterised by a division of labour and the development of hitherto

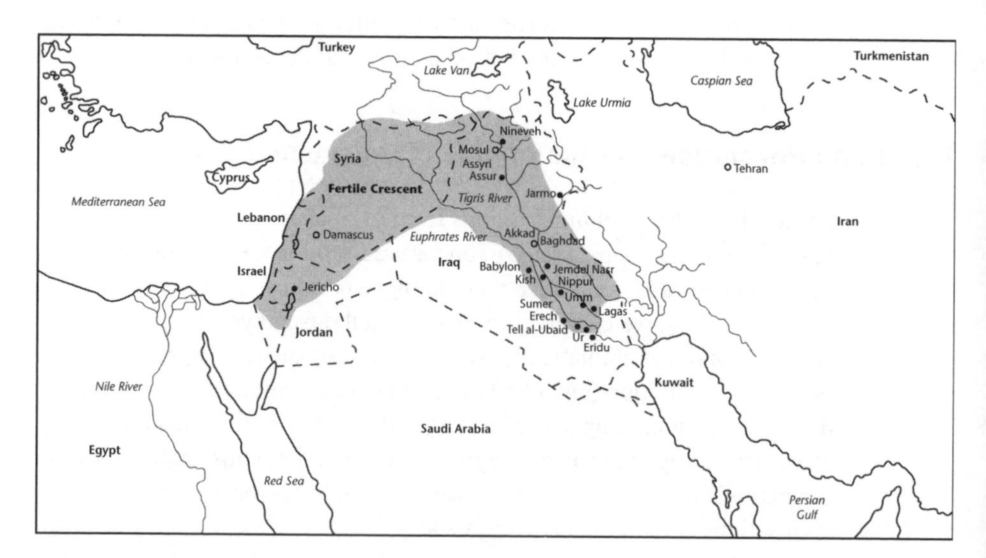

Figure 2.1 The origins of urbanism in the Middle East

Source: Pacione (2005)

unknown degrees of social hierarchy. Indeed, 'for archaeologists and historians the most meaningful difference between a village and a city has nothing to do with size; it is instead a measure of social and economic differentiation within the communities' (Reader 2005: 16). The shift from subsistence agriculture to surplus agricultural production made it possible for a significant number of people to specialise in the production of crafts, the arts of war, the cultivation of philosophy and activities related to trade. As a result, cities became sites of social and technological innovation, giving rise to improvements in irrigation, transportation and metallurgy. A rising demand for inputs into new productive activities stimulated long-distance trade. It was also in cities that writing first emerged, which facilitated the development of mathematical sciences and made bureaucracy and public administration possible. The combined effect of these innovations was a radical transformation in the organisation of society, ultimately resulting in the rise of the first civilisations and empires. As Bairoch notes, 'the city appears to gather together all of the factors conducive to sociotechnical advance' (Bairoch 1988: 96). The standard narrative, then, is that agriculture made cities possible, which in turn gave rise to socio-economic differentiation, socio-technical innovation, and (eventually) the first civilisations.

Edward Soja (2000), and Jane Jacobs (1969) before him, have challenged this narrative by suggesting (based on archaeological evidence) that cities were, in fact, both the impetus for, and the incubators of, the innovations that made the production of an agricultural surplus possible. Excavation sites at Jericho (in present-day Palestine) and Çatal Hüyük (in present-day Turkey) have revealed evidence of large, dense, permanent settlements established approximately ten thousand years ago – about the same time that the Neolithic Revolution took place. The inhabitants of Jericho are thought to be the first in the world to live sedentary lives, although they were hunter-gatherers. Çatal Hüyük, a larger and denser settlement, also displays evidence of a hunter-gatherer lifestyle, but there is some evidence of early agriculture in and around the city. Soja argues:

> Jericho and Çatal Hüyük represent a revolutionary leap in the social and spatial scale of human societies and culture . . . The stimulating interdependencies and cultural conventions created by socio-spatial agglomeration – moving closer together – were the key organising features or motor forces driving virtually everything that followed.
>
> (Soja 2000: 46)

So which came first, farming or cities? The answer (given the available evidence) depends upon one's definitions of 'city' and 'urbanism'. Childe,

Mumford and Bairoch dismiss Jericho and Çatal Hüyük as pre-urban towns on the grounds that there is little evidence of socio-economic differentiation in these settlements given their reliance on hunting and gathering, and that they were relatively small in comparison to the early Sumerian cities. Nevertheless Soja challenges us to consider the possibility that these settlements were indeed cities on the grounds that they exhibit the 'socio-spatial agglomeration' that characterises urbanism. There is no disagreement on the facts, but rather which aspect of urbanism these scholars choose to privilege.

In making his case, Soja places emphasis on what he calls *synekism*, a respelling of the word synoecism (pronounced sin-ee-sism), which is used in the archaeological and historical literature on ancient cities. Synoecism is derived from the ancient Greek *synoikismos*, which literally means the 'conditions arising from dwelling together in one house' (Soja 2000: 12). Similarly, synekism refers to

> the economic and ecological interdependencies and the creative – as well as occasionally destructive – synergisms that arise from the purposeful clustering and collective cohabitation of people in space, in a 'home' habitat.
>
> (Soja 2000: 12)

The power of synekism – also referred to by urban scholars variously as density, propinquity and proximity – can be conceptualised as a kind of socio-spatial force or stimulus unique to urban agglomerations.

By overlooking the possible role of synekism in Jericho and Çatal Hüyük, the standard narrative sees the rise cities as a by-product of the Neolithic Revolution as opposed to a contributing factor in this critical juncture in human history. It is unlikely that we will ever know the true sequence of events, but the controversy highlights two essential aspects of urbanism – synekism and socio-economic differentiation – that are driving forces behind socio-technical change.

From the birth of the city to the rise of nation-states

The cities that emerged in ancient Mesopotamia were not merely cities, but rather city-states – sovereign urban agglomerations in control of their immediate hinterlands. Their socio-political order was characterised by divine kingship, and the demands of acquiring, defending and distributing an agricultural surplus inspired the evolution of what we recognise today as basic state functions, such as taxation, military conscription, policing and bureaucratic administration. Religion and ritual played an important role in justifying this new order and temples served as sites of worship, public administration and (importantly) granaries (Childe 1950).

Ur was the first and paradigmatic Sumerian city-state, with an urban population of about 24,000 people (circa 2800 BCE) governing and extracting a surplus from some 500,000 farmers in the areas around the city (Bairoch 1988: 26). The city was built on a *tell*, or raised mound, and was surrounded by an oval wall oriented on a north–south axis, located close to the outlets of the Tigris and Euphrates rivers. Its urban centre was dominated by a ziggurat some five storeys tall, which contained administrative offices, storage facilities and, at the uppermost level, a shrine. It was a physical manifestation of the centralisation of economic, political, social and spiritual power that would be replicated on increasingly larger scales across the network of city-states that proliferated in Sumer. The Tower of Babel, the ziggurat dominating the skyline of Babylon as it ascended in the ancient world, was some 270 feet (27 storeys) tall.

Although the Mesopotamian city-states provide the earliest evidence of the development of social, economic and political structures that would later become the organisational foundations of civilisations, empires and nation-states, urban history is not linear. City-states and urban-based empires with very different characteristics emerged subsequently and independently across the world in the centuries that followed (Figure 2.2).

The rich urban history of India began east of Mesopotamia in the Indus Valley (present-day Pakistan) where the cities of Harappa and Mohenjo-Daro are thought to have been built in the second millennium BCE. They were home to up to 40,000 people, had functioning sewers and appear to have been built on a regular plan (Bairoch 1988: 39–40). They were ruled by a single 'priest-king', and appear to have traded with the Sumerian city-states, but their construction does not appear to have been influenced by the birth of cities in Mesopotamia (Pacione 2005: 44). Delhi, the modern capital of India, has been the site of seven cities over the past two thousand years.

Cities also emerged independently in China as early as 1600 BCE – some evidence suggests even earlier – in the Wei River Valley during the Shang dynasty. Similarly to Mesopotamian cities, urban centres in China served important spiritual, economic and political-administrative functions. At its peak, Chang'an (capital of the Sui and Tang dynasties) may have been

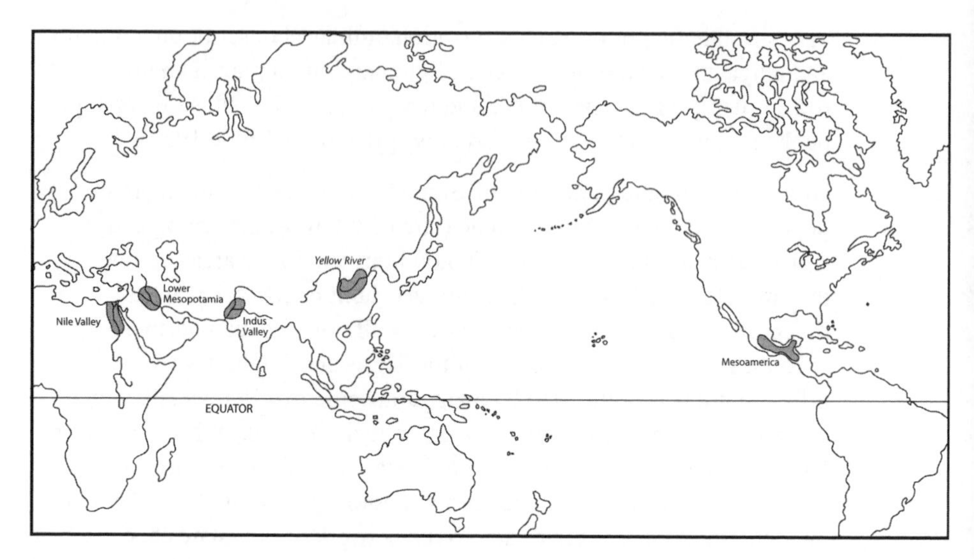

Figure 2.2 Other early urban centres

Source: Pacione (2005)

home to 1 million inhabitants. It was formally planned and carefully regulated, its morphology and regimented daily life reflecting the rigid political hierarchy of the time. The following imperial age saw a dramatic expansion of Chinese territory and population growth, rendering centralised governance of the daily lives of urban residents difficult. Local elites organised into a range of civic associations and increasingly provided the framework for local governance. As the vast Chinese empire was consolidated through tumultuous episodes of expansion and retraction, cities served critical functions as cultural, intellectual, economic and political hubs (Friedmann 2005).

Not long after the rise of Mesopotamian cities, city-states also appeared in Phoenicia along the eastern shores of the Mediterranean (present-day Lebanon, Syria and Israel). These cities were among the first truly commercial or merchant cities in the world, trading extensively throughout the Mediterranean. They served as hubs of inter-regional commerce making extensive use of maritime trade as far West as the Atlantic Ocean. There is evidence of a significant shipbuilding industry, as well as the production and export of glassware and dyed cloth. Unlike their Sumerian cousins, they were not ruled by kings, but by councils of elders drawn largely from the merchant class (Bairoch 1988: 30; Parker 2004). Although they never reached the size of the Sumerian city-states, their function as centres of commerce – indeed their reliance on commerce – is a recurrent theme in urban history and one that later played an important role the evolution of nation-states.

North across the Mediterranean, the Greek polis was a particularly influential urban form, serving as a centre of cultural innovation that generated a canon of intellectual, architectural and artistic works that continue to inform contemporary social, political, scientific and aesthetic endeavours (Hall 1998). Rome, which began as an Etruscan colony, became an independent city-state in the sixth century BCE and grew into one of history's greatest empires, reaching an unprecedented size (some say 1 million inhabitants by AD 100), compelling its rulers to 'devise complex systems of international food supplies, to grapple successfully with long-distance delivery of water and with complex systems of waste disposal, even to formulate rules of urban traffic management' (Hall 1998: 621). The Romans were arguably the first masters of urban planning, recording their strict imperial order in the geometry of new cities with grid-iron planning. These cities were the urban nodes of a networked empire, connected to the core by an impressive road network. The ancient infrastructure of the empire can still be found in cities across Europe, Northern Africa and Western Asia.

In the Americas, great cities were established in Mesoamerica and the Andean mountain range long before Spanish colonisers arrived. The Aztec capital of Tenochtitlán, with its sparkling pyramids, famously inspired awe in Fernando Cortés and his men as they entered the Valley of Mexico for the first time; the Mayan city of Tikal covered more than 123 square kilometres and was home to some 45,000 people circa AD 550, and the Incan city of Cuzco, which reached a peak population of perhaps 300,000 inhabitants, was known as the 'city of bureaucrats', who managed an extensive road network and 170 administrative satellites (Butterworth and Chance 1981).

While the urban history of Africa has received less attention from archaeologists, there is evidence of cities in the kingdoms of Ghana and Cush before the turn of the millennium, as well as large urban settlements such as Aksum (circa AD 100–600) in the Horn of Africa and Great Zimbabwe (circa AD 1000–1500) in Southern Africa serving as important centres of regional governance and inter-regional commerce (Anderson and Rathbone 2000). One of the more enigmatic discoveries has been the remains of Jenne-Jeno in present-day Mali, which was established in the third century BCE and had perhaps 20,000 residents by AD 800. There is evidence from the site of the economic specialisation typical in all cities, but unlike so many other ancient cities, there is no evidence of political centralisation or social stratification (Freund 2007; McIntosh and McIntosh 1981).

Everywhere that cities 'crystallised', to use Mumford's (1961) term, they served as nuclei around which new socio-political orders revolved, as centres of trade and incubators of new technologies. Their political influence often extended far beyond their immediate hinterlands, and their socio-technical innovations were diffused through trade. For thousands of years, cities reigned supreme as 'proto-states', sometimes competing with one another, sometimes forming strategic federations. But in medieval Europe, a very particular kind of collaboration emerged between urban-based merchants and territorial monarchs that culminated in the rise of nation-states.

Charles Tilly (1994: 6) has argued that 'the variable distribution of cities and systems of cities by region and era significantly and independently constrained the multiple paths of state formation' in Europe. He observed that different kinds of states emerged in regions with few cities as opposed to densely urban ones, that organisationally advanced urban centres played a significant role in national politics, and that urban merchants and financiers played an integral part in the financing and provisioning of new states through their control of capital and markets. According to Tilly (1994), the rise of the territorially defined and centrally governed nation-states that we are accustomed to nowadays was essentially the product of a strategic collaboration between urban-based capital and rural-based coercion. Urban-based merchants and financiers were specialists in acquiring, managing and deploying capital, as they relied heavily on trade to accumulate wealth. By contrast, feudal landlords and petty depots were specialists in the use of coercion, or armed force, and amassed their wealth through taxing peasants.

Before 1500 or so, monarchs maintained armies drawn from their own subjects who owed them personal service. But as the frequency and intensity of territorial conflicts increased in the period 1500–1700, it became necessary to employ mercenaries, which in turn required capital. Monarchs turned to wealthy urbanites to finance their wars, but this support carried a cost. The urban classes demanded certain securities in return, which led to a bargaining process. In Tilly's own words:

> In Europe before 1800 or so, most important changes in state structure stemmed from rulers' efforts to acquire the requisites of war, from

> resistance to those efforts, and from bargains that ended – or at least mitigated – that resistance. Courts, treasuries, representative assemblies, central administrations, fiscal structures, and much more formed and reformed in response to the creation of military force, the pursuit of war, and the payment of its costs.
>
> (Tilly 1994: 10)

In effect, the institutions developed over time in cities to accumulate and manage financial resources, and the institutions developed by monarchs to ensure coercive dominance over their populations and territories came together to create a powerful political unit. In the nineteenth century, rulers

> continued to bargain with capitalists and other classes for revenues, manpower, and the necessities of war. Bargaining, in its turn, created numerous new claims on the state: pensions, payments to the poor, public education, city planning, and much more. In the process, states changed from magnified war machines into multipurpose organisations.
>
> (Tilly 1994: 9)

The result was the political form of the nation-state, which remains today the pillar of the international political-economic system, despite the forces of globalisation. The gradual consolidation and solidification of nation-states in Europe and across the globe continued well into the twentieth century. At the same time, beginning in the latter half of the eighteenth century and continuing to the present day, a critical tripartite transition began: a demographic transition, an economic transition and an urban transition.

The demographic transition and the Industrial Revolution

Despite the fact that cities existed for thousands of years, truly urbanised societies did not emerge until the twentieth century. Cities had grown in both number and size over the previous millennia, and incremental improvements in agricultural productivity had managed to keep pace with growing urban populations, but cities still remained small by modern standards. As Figure 2.3 demonstrates, the world's population remained overwhelmingly rural until the end of the nineteenth century, when the balance between rural and urban settlements began to shift rapidly in Europe and North America (and slightly later in Japan). This sudden acceleration in urbanisation coincides with both the first demographic

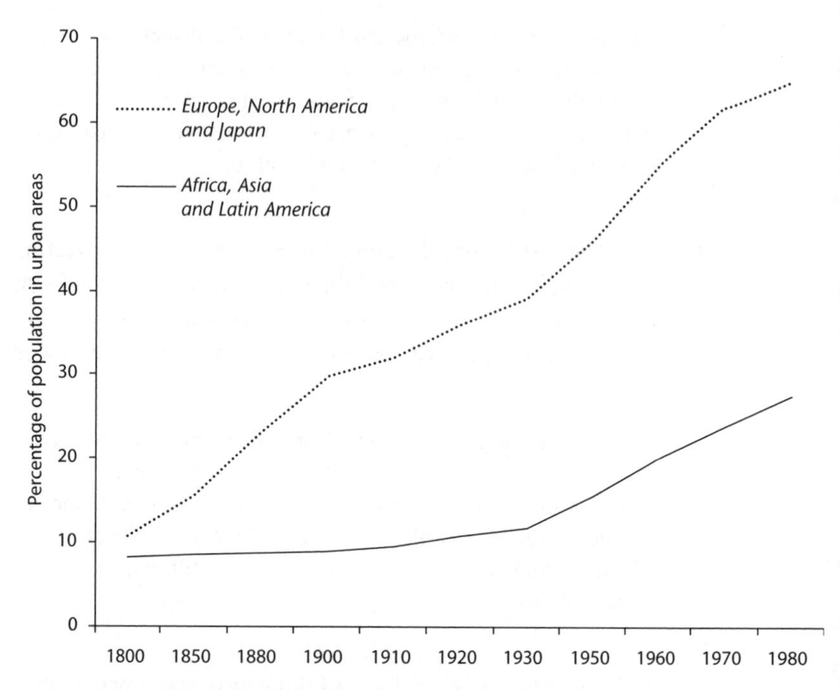

Figure 2.3 Historical urbanisation trends

Source: Bairoch (1988)

transition and the early phases of the Industrial Revolution, and was both a product and a catalyst for these important transitions.

Before the twentieth century, cities everywhere were demographic 'sinks'. High population densities in cities coupled with poor sanitation systems and poor hygiene created an ideal environment for infectious diseases to flourish. High mortality rates effectively placed a 'ceiling' on urban growth, and cities depended on in-migration to reproduce themselves (Dyson 2001: 8). In the eighteenth century, however, mortality rates began to fall, heralding the beginning of what demographers refer to as the 'demographic transition'. A demographic transition occurs in a population when mortality rates and fertility rates both decline. But they do not decline simultaneously. Generally speaking, mortality rates fall first, and only after time do fertility rates fall, creating a spurt of rapid population growth in the interval. The demographic transition (which is a universal phenomenon, although its onset has differed significantly across regions) marks a shift from a wasteful cycle of reproduction in which a high number of both births and deaths occur, to a more economical one where fewer are born but are more likely to survive and live longer lives (Livi-Bacci 2001).

The first demographic transition began in Europe in the eighteenth century with a gradual decline in mortality due, perhaps, to the availability of a more stable supply of food thanks to rapid innovations in agriculture and improved hygiene (Livi-Bacci 2001). In the eighteenth century, better infrastructure and steady improvements in medicine made possible the control of infectious diseases, contributing to further mortality decline and inverting the negative balance between fertility and mortality in cities, thereby removing the ceiling on urban growth. But the demographic transition was not completed until well into the Industrial Revolution.

The Industrial Revolution gathered pace in Britain in the late nineteenth century, following a period of rapid improvements in agricultural production (referred to as the British agricultural revolution), which enhanced the surplus available to feed non-agricultural populations – possibly contributing to the initial decline in mortality rates noted above. Although there is some evidence that urban demand provided an incentive for the innovations that made this possible, it is not clear that cities played a defining role in this productivity boom. Nor did cities play a decisive role in igniting the early stages of the Industrial Revolution (Bairoch 1988: 331): levels of urbanisation in Europe at the time are negatively correlated with growth rates in the twentieth century (Bairoch 1988: 261). There is no doubt, however, that industrial transformation was largely responsible for the increased pace of world urbanisation through the nineteenth and twentieth centuries, that it significantly influenced patterns of urbanisation and urban development, and that in time, cities contributed to the acceleration and diffusion of the Industrial Revolution.

While pre-industrial cities were largely centres of political administration, commerce, and small-scale production of specialised goods, industrial cities were centres of manufacturing, which required immense amounts of energy and labour (Bairoch 1988: 269). In the early days of the Industrial Revolution in Britain, it was small towns and rural areas that contributed the most of both. This can largely be explained by the technology of the time: machines were driven by water-power and the iron and steel industries depended heavily on coal. In Britain, this resulted in a restructuring of the urban hierarchy as firms chose to locate close to these resources (and the cheaper labour available in these areas), thereby stimulating employment and construction outside of the major urban centres of the day (Bairoch 1988: 331). Later, the invention of the steam engine released industry from a reliance on water-power, permitting the construction of the first great factories with purpose-built housing for workers close by. The development of railway networks significantly reduced the costs of transporting food, fuel and raw materials (not to mention people), further stimulating city growth (Bairoch 1988: 276–278). Improvements in agricultural productivity, due to better organisation and technology, facilitated the shift of the working population away from agriculture and towards industry. The cumulative effect of these changes on urbanisation was profound. Bairoch notes that the United Kingdom alone was responsible for 35 per cent of all urban population growth in the 'developed world' (i.e. Europe and North America) between 1800 and 1850, and between 1800 and 1900 the urban population of the United Kingdom grew from 3.1 million to 27.8 million (Bairoch 1988: 290).

If cities did not kick-start the demographic transition or the Industrial Revolution, they certainly contributed to their acceleration and diffusion. The dramatic social, cultural and economic changes associated with the Industrial Revolution and increased urban habitation contributed to a subsequent decline in fertility rates: as the costs of having children rose along with the benefits of investing more in their health and education, fertility rates slowly declined. This decline began in cities, but eventually spread to rural areas and to less economically advanced nations (Galor 2005; Livi-Bacci 2001: 94).

Just as cities contributed to the completion of the demographic transition through the effects of urbanism on fertility rates, cities catalysed the Industrial Revolution, spurring innovation and the diffusion of technology. There is ample historical evidence indicating that innovation was directly correlated with levels of urbanisation and with city-size once the Industrial Revolution was underway in Britain, Europe and North America (Bairoch 1988: 323–325). New ideas travel fast in cities, bolstered by better access to education, improved social mobility, and access to mass media, all of which facilitate the production, diffusion and cross-fertilisation of knowledge and technologies (Bairoch 1988: 327). Furthermore, the enhanced division of labour characteristic of urban economies opens up new possibilities for productive innovations by specialists in particular trades or industries.

The combined forces of the demographic transition and the Industrial Revolution set in motion an urban transition that continues to this day. In turn, burgeoning urban populations adopted new reproductive habits and catalysed the economic transformation that radically restructured the way we live today. Figure 2.4 illustrates the relationships driving this tripartite

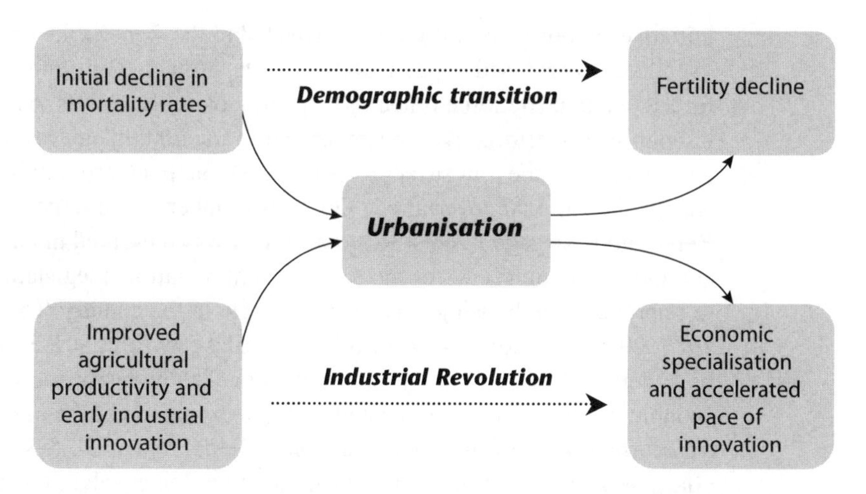

Figure 2.4 Interaction between demographic, economic and urban transitions

transition. But the urban transition engendered by the demographic transition and the rise of industry was accompanied by myriad problems. Cities were unprepared for such a massive shift in the balance of human settlements. In the 33 years between 1847 and 1880, for example, London grew from 2 million to 5 million inhabitants, and the human consequences were severe (Bairoch 1988: 285). Friedrich Engels (1945) described the conditions of London's slums in the middle of the nineteenth century thus:

> The streets are generally unpaved, rough, dirty, filled with vegetable and animal refuse, without sewers or gutters, but supplied with foul, stagnant pools instead. Moreover, ventilation is impeded by the bad, confused method of building of the whole quarter, and since many human beings here live crowded into a small space, the atmosphere that prevails in these working-men's quarters may readily be imagined.
>
> (Engels 1945)

Such conditions were hardly unique to English cities; across Europe and North America similar squalor could be found. In some cases, wretched living conditions sparked political resistance that resulted in significant changes in public policy. For example, around the turn of the twentieth century, Glasgow was overflowing with labourers working in the shipbuilding and naval industries. Organised labour groups fought for over a decade for legislation to improve housing conditions, but were repeatedly thwarted by powerful landlords. However, a grassroots initiative supported by local housewives led to the Rent Strike of May 1915. Strikers refused to pay increased rents, protected one another from forced eviction (by force if necessary) and took to the streets to support Labour reform efforts. By November there were 20,000 city residents participating in the rent strike, forcing the Municipal Corporation to engage the national government. On 25 November of the same year a Rents and Mortgage Interest Restriction Bill was presented in Parliament, and the following years saw the development of national legislation to improve urban housing conditions throughout the country (Castells 1983: 29–30). Theoretically speaking, the Glasgow Rent Strike can be understood as a case of synekism at work. The intensification of economic specialisation associated with the expansion of the industrial transformation drew increasing numbers of labourers into cities such as Glasgow, and the conditions in which they found themselves collectively sparked the formation of organisations and networks that ultimately transformed the political-economic institutions governing their housing conditions.

The power of the city to foment political reform has been observed by many scholars. Figures 2.5a and 2.5b from Dyson (2001) illustrate the positive relationship over time between urbanisation and democratisation across the world. While not claiming that urbanisation is the decisive factor in bringing about democracy, Dyson (2001: 17) notes that

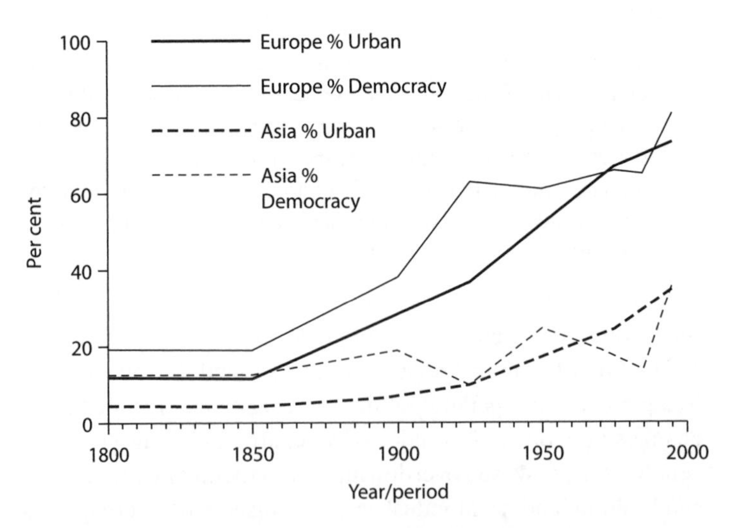

Figure 2.5a Urbanisation and democratisation, Europe and Asia

Source: Dyson (2001)

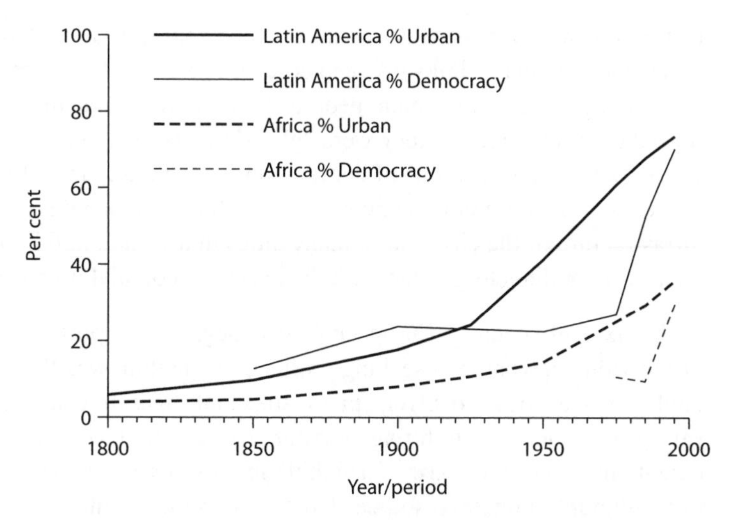

Figure 2.5b Urbanisation and democratisation, Latin America and Africa

Source: Dyson (2001)

'Urbanisation focuses attention on the distribution of political power in society, so helping to bring about the rise of modern democracy'. Similarly, Borja and Castells (1997: 251, 246) argue that cities are 'privileged places for democratic innovation', providing 'a chance to build a democracy of proximity, of participation by all in the management of public affairs'. The urban milieu is clearly conducive to the organisation of political action (see Chapter 7), and may very well be partialy responsible for the spread of democracy worldwide. At the same time, cities can and have been instruments of imperial control and colonial exploitation.

Colonial urbanism

Beginning in the sixteenth century European powers began to expand their spheres of influence, establishing settlements and imperial outposts in the Americas, Asia, Africa and across the Pacific. Over the next few hundred years, mercantilist enterprise gave way to imperial conquest and colonialism. Colonialism – defined as the political control of a people and territory by a foreign state (Bernstein 2000) – was characterised by political domination, social control, racism and exploitative economic relations, all of which have left their mark on the development trajectories of former colonies. Throughout the colonial era 'the city and its institutions were a major instrument of colonisation' (King 1990: 29), serving as 'the major links between core and peripheral economies . . . articulating the flow of capital, people, commodities, and culture that flowed between them . . . they were "global pivots of change"' (King 1990: 7). Colonial relations of domination were literally inscribed in the form of the built environment: in the architecture, institutions and infrastructure of the city. Today, many cities in low- and middle-income countries continue to grapple with the legacy of colonial urbanism.

In the sixteenth century, European powers began to support the exploits of individual adventurers seeking profits in the natural wealth of foreign lands, such as gold and silver, spices, silks and sugar. In general, early European settlements in foreign territories were small; companies essentially established bases for linking into existing local trade networks or tapping into mineral resources (such as silver mines in South America). But as profits grew, so too did the need to establish more permanent arrangements with warehouse facilities and military garrisons for protection.

In the sixteenth century, conquistadors received instructions from the King of Spain (first Ferdinand and then Philip II) concerning the appropriate locations and forms of cities to be built in Central and South America. In an interesting case of continuity in urban form, the Spanish kings based their urban planning instructions on Roman traditions, apparently drawing on the works of Vetruvius – a famous Roman architect and engineer who lived during the first century BCE. The location and orientation of cities were dictated, as well as the geometry of plazas, street plans, and the placement of cathedrals, town halls and other administrative buildings (Stanislawski 1947). After Hernán Cortés destroyed Tenochtitlán and most of its inhabitants, he rebuilt it in accordance with the wishes of the Spanish king and in line with Vetruvius' orderly vision (see Plate 2). This was an urban plan that was repeated across Central and South America and can be observed at the present time in the older cities of Latin America.

While most Latin American countries achieved independence in the first two decades of the nineteenth century, their political and economic trajectories over the subsequent two centuries were heavily influenced by the colonial episode. The governments that took over at independence were generally comprised of the elite decedents of European colonisers who had amassed great fortunes in mining and plantation agriculture. The indigenous populations that had managed to survive the initial colonial

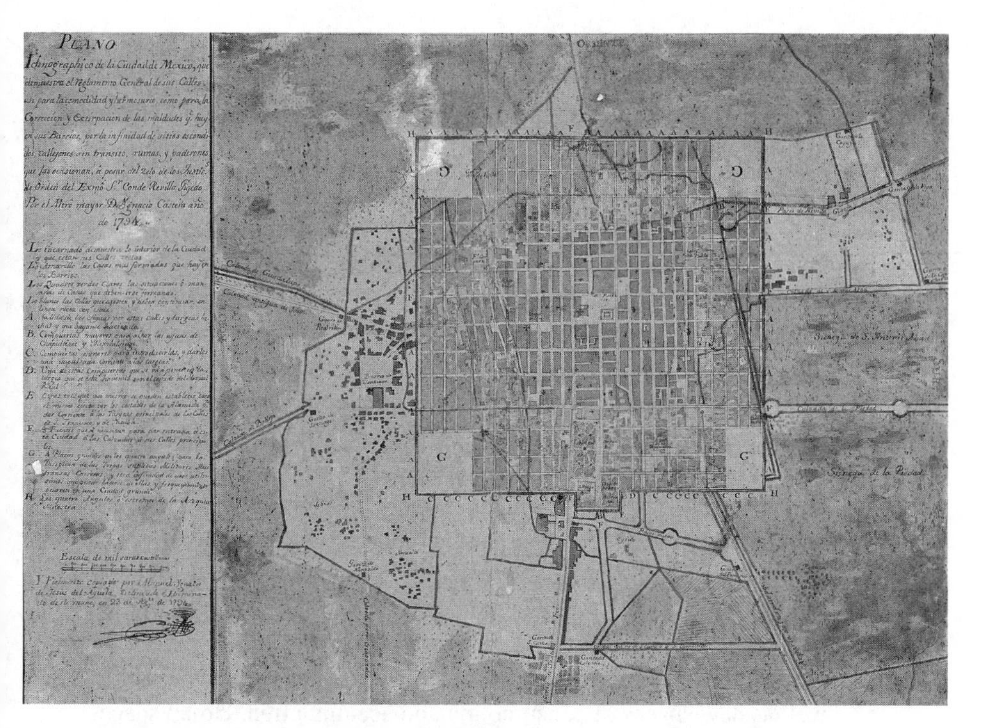

Plate 2 Plan of Mexico City in 1794 by Manuel Ignacio de Jesus del Aguila

Source: Library of Congress, Geography and Map Division

onslaught of military conquest and European diseases were relegated to subsistence agriculture, or forced to work in mines or on large farms. Powerful landlords continued to dominate politics, perpetuating the gross inequalities established in the colonial era. Even today, legacies of economic inequality and racial discrimination remain.

In other regions, the Dutch, French, British and Portuguese began establishing small commercial and residential enclaves, adapting architectural forms and planning norms from their homelands to local climates and available materials. While in some cases initial colonial encounters led to the destruction of existing populations and urban centres (as in Spanish America), the early phases of colonialism generally left existing cities and urban hierarchies intact (Drakakis-Smith 2000). In the late nineteenth and early twentieth centuries, however, interests in the colonies began to change, and so too did the role of cities. War in Europe reduced the availability of venture capital to be spent on colonial enterprises, and many of the companies that spearheaded mercantilist ventures were forced to hand over control of their overseas possessions to state governments. Not long after, the Industrial Revolution kicked into gear in parts of Europe, increasing the demand for food and raw materials to supply bourgeoning urban populations and new industries. Colonies, in turn, played an important new role in the economic expansion of Europe, providing both raw inputs for production, as well as markets for manufactured goods (Drakakis-Smith 2000).

Over the course of the late nineteenth century, European powers significantly expanded their presence in what had previously been small colonies through the active acquisition of territory and investments in infrastructure. This industrial phase of colonialism has received much attention from dependency theorists, who see the history of colonialism as one of integrating Latin America, Africa and Asia into a capitalist world system in which peripheral countries produce primary commodities for core countries while absorbing the excess productive capacity of core countries by purchasing manufactured goods. For example, by 1931 two-thirds of British exports went to its colonies, and just less than half of all imports came from its colonies (Christopher 1988, cited in King 1990: 5). In this context of increasing economic interdependence, colonial cities served as political, administrative and military nodes of empire, as well as new spaces of consumption and accumulation. Urban spaces were transformed through the collision of cultures, the form of political domination and the technologies imported from industrialising powers (King 1976).

Within cities, racial and ethnic segregation became a standard feature of colonial urban morphologies. New planning concepts imported from Europe were used to create 'sanitary' districts in colonial urban settlements, and reinforce notions of European superiority. Across many British colonies, the central districts of European-designed cities were usually surrounded by a green belt, a *cordon sanitaire*, within which no colonial subject was allowed to live. In certain Nigerian towns, the green zone was to be at least 440 yards (402.3 metres) wide since it was then believed that this was further than a malarial mosquito could fly (Hardoy and Satterthwaite 1989: 20). In Lusaka, Zambia, African inhabitants were made to live in village-like compounds within the city (see Box 2.1). In Delhi, India, urban planning driven by sanitary concerns in the late nineteenth century was accompanied by 'nuisance laws' to regulate the ablution and sanitation activities of urban residents. One could be found guilty of public nuisance for any

> negligent act likely to spread infection of disease dangerous to life. . . . [any] malignant act likely to spread infection of disease dangerous to

Box 2.1

Eric Dutton's vision for colonial Lusaka

Lusaka was chosen as the site of a new administrative capital for Northern Rhodesia (now Zambia) because of its location halfway between the settler-dominated city of Livingstone in the south and the economic lifeblood of the Copperbelt in the north. At the time, Lusaka was just a railway watering station on the line linking the two, and as such was built almost entirely from scratch. Eric Dutton, the assistant chief secretary at the time, was hugely influential in terms of how the colonial capital was constructed.

Lusaka was in many ways designed as an archetypal 'garden city', with rows of trees and hedges strategically planted to divide compounds, hide African areas from the traveller's eye and generally – in Dutton's own words – conceal 'a multitude of sins'. The African compounds were based on elements of the traditional African village, carefully placed within the larger plan of Lusaka. The servants' compounds consisted of several units of four small huts looking in on themselves, emphasising a clear notion of inside and outside and reinforcing the idea that African residents belonged in one clearly defined section of their 'village', and should concern themselves only with that part of it. There were no central gathering places, and strategically located hedges were used to spatially segregate units and prevent the emergence of a broad-based – and potentially political – urban consciousness. These compounds were also built on low floodable land looking up at the Ridgeway, the location of the Government House, which was located so as to be visible to the African population as frequently and widely as possible.

Other than these 'showpiece' compounds, little regard was paid to Africans in terms of the way Lusaka was designed, and most Africans ended up in the expanding unplanned areas of 'Old Lusaka'. In effect, these settlements undermined Dutton's planning, and as Myers (2003) highlights, it is the way in which the urban majority, excluded from planning processes, reframed colonial cities that ultimately had the more significant impact on their evolution. Indeed, 'Even as the colonial regime seemed to try and make Lusaka readable like a book . . . the poetry of that book was being rewritten from underneath' (Myers 2003: 56).

Source: Myers (2003)

> life . . . fouling water of public spring or reservoir . . . making atmosphere noxious to health . . . and negligent conduct with respect to poisonous substance.
>
> (from the penal code of 1862, cited in Sharan 2006: 4906)

Street traders, bathers and fishermen could all be guilty of nuisance for various reasons. The physical organisation of city-space and the regulation of activity in that space were complementary strategies of control and segregation in Delhi.

Almost without exception, the economies of colonial possessions were not encouraged to diversify and grow internally. Indeed, they were actively discouraged from doing so. Colonial powers ruthlessly exploited the natural resources and peoples under their control, extracting what profit they could without concern for the human or environmental consequences. This extractive economic system was reflected in the restructuring of urban hierarchies. Colonial capitals and ports grew disproportionately with their political and economic power derived from direct links with the colonial metropole. Pre-colonial urban settlements were often destroyed or co-opted. In Sub-Saharan Africa, pre-colonial urban centres were generally found inland as most trade was conducted overland as opposed to by sea. The building of colonial ports on the western, southern and eastern coasts of Africa inverted the traditional urban hierarchies, drawing power and economic activity towards the colonial cities on the coasts, or along critical transport lines (e.g. Nairobi).

This shift in urban hierarchies was facilitated by the pattern of infrastructure development in this industrial phase. Linked infrastructure, such as railways and telegraph lines, were extended more deeply into colonial territories, but these networks 'were designed mainly to evacuate exports. There were few lateral or intercolonial links, and little attempt was made to use railways and roads as a stimulus to internal exchange' (Hopkins 1973, quoted in Graham and Marvin 2001: 84). In effect, infrastructure was designed to funnel all goods to a port.

In the latter phase of colonialism, there was a significant boost in urban infrastructural development, reflecting the modernist ideals and moral discomfort with exploitation emanating from Europe and the demands associated with attempts to improve local productivity through land reforms and mechanisation. Urban migration accelerated and cities began to swell with indigenous populations (Drakakis-Smith 2000). In response, many colonial governors felt the need for more extensive urban planning. The 'Garden City' movement in England influenced British colonists in particular, who used urban planning to re-enforce segregation and reflect, in the built environment, their perception of their own superiority (Graham and Marvin 2001: 82). Cities where settler colonialism was a feature, such as Harare, Nairobi and Lusaka, share a similar morphology in the strict separation of administrative, commercial, industrial and residential space, a separation typical of the modernist concepts being promoted in Europe at the time (O'Conner 1983: 199). In Nairobi, residential areas were further segregated by race, with white Europeans, Asians and Africans living in separate quarters of the city. Some colonial governments – such as in Northern Rhodesia, or present-day Zambia – increased investment in indigenous quarters in an attempt to improve the stability and productivity of labour to support industrialisation and 'civilise' native populations (Mabogunje 1990: 127; see also Heisler 1971). These investments had consequences for colonial rulers. The urban milieu was home to an increasingly educated and disgruntled urban African population who found new ways of organising resistance:

> Associational life . . . was getting richer and more varied, built around occupational and residential affinities, around connections to common regions of origin, around churches or mosques or indigenous religious institutions, around mutual aid needs of various sorts, around the new forms of music and artistic creation. Most important was the mix of urban-born and migrant youth, a category marked, as Rémy Bazenguissa-Ganga puts it, by its 'availability' – a vibrant, volatile force that could be channelled in different directions.
>
> (Cooper 2002: 34–35)

Resistance movements emerged and spread across the remaining colonial empires in Africa, Asia and the Middle East, leading to a long period of decolonisation in the aftermath of the Second World War. And although independence struggles were not a strictly urban phenomenon, the urban milieu provided an essential space for the organisation of resistance, for the acquisition of essential resources and as information and communication hubs.

Despite dramatic differences in the nature of colonial experience and timing of independence movements across the world, most former colonies were left with infrastructure designed to funnel goods abroad instead of encourage domestic circulation, with national boundaries that did not reflect pre-colonial political geographies, with unbalanced urban hierarchies, with cities designed to segregate, and with a variety of institutions – such as property regimes and regulatory frameworks – unfit for inclusive development.

Into the first urban century

The end of the Second World War, and the wave of decolonisation that it spawned, coincided with an acceleration of international economic integration and the commencement of the Cold War. At first, post-colonial governments, taken with modernisation theory and intent upon establishing a sense of national identity and ownership, embarked upon large-scale infrastructure and economic development projects to redress underinvestment during the colonial era. Airports, multi-lane highways, power stations, public housing and monuments to the leaders of the independence struggle were built (Graham and Marvin 2001). But these extensive investments failed to generate the benefits imagined by their designers. A post-war boom in the global economy turned into a global economic downturn in the early 1970s, undoing what little progress had been made in previous decades, and sending many low-income countries into heavy debt. The political legacies of colonialism began to reveal themselves, thwarting early nationalist ambitions by exacerbating conflicts within and between nations. In Africa (and arguably elsewhere), colonial governments had behaved as gatekeeper states standing 'astride the intersection of the colonial territory and the outside world. Their main source of revenue was duties on goods that entered and left the ports' (Cooper 2002: 5). Post-colonial governments, inheriting economies oriented toward extraction,

> realized early on that their own interests were served by the same strategy of gatekeeping that had served the colonial state before World War II: limited channels for achievement that officials controlled were less risky than broad ones which could become nuclei for opposition. But the post-colonial gatekeeper state, lacking the external capacity of its predecessor, was a vulnerable state, not a strong one.
>
> (Cooper 2002: 5)

Corruption, coups, conflict and outright war became regular features of the post-colonial era. At the same time, many former colonies found themselves caught up in the power struggle of the Cold War, with the United States and the Soviet Union (among others) intervening in the political and military of affairs of nations struggling to consolidate political order and stimulate economic development. In many cases, these Cold War enemies propped up dictators and financed proxy wars – particularly in Asia and Africa – with devastating consequences. The brief spells of hope that had attended the early days of independence quickly faded.

Through all of this, cities continued to serve as nodes in the international political economy and nowadays many cities in low- and middle-income countries remain more economically and politically bound to cities in Europe, North America or East Asia than to their own hinterlands. The legacy of extractive economic relations has been difficult to overcome, with many countries continuing to rely on exports of primary commodities (such as agricultural products and mineral resources) that perpetuate economic vulnerability and dependency. Although the political bonds of colonialism have been broken, many scholars argue that contemporary economic relations between industrialised and less industrialised nations are a form of 'neo-colonialism' exercised through the organisation of the global economy; through transnational corporations (TNCs) and multilateral agencies such as the World Bank and IMF, which exert enormous influence in low- and middle-income countries.

Within cities, colonially imposed 'laws, norms and codes governing housing, building and planning . . . remain today, largely unchanged . . . Not surprisingly, they are very poorly suited to [independent nations] where rapid urban growth and increasingly urban societies have become the norm' (Hardoy and Satterthwaite 1989: 20). While well-serviced urban quarters once served to segregate Europeans from indigenous populations, they now serve to insulate domestic political and business elites, who have maintained a high standard of living by capitalising on their role as gatekeepers while allowing the formerly 'indigenous' quarters of their cities to deteriorate (Mabogunje 1990: 141).

Against this background, cities in low- and middle-income countries have been growing at historically unprecedented rates. While the demographic transition is effectively complete in most industrialised nations, it is still very much underway in the rest of the world. Figure 2.6 presents the trends in population growth rates in the North (Europe and North America) and South (Africa, Asia and Latin America), illustrating the early and gradual demographic transition in the North, and the later and more rapid transition in the South. Livi-Bacci (2001) explains the dramatic difference in the timing and rate of the demographic transition in the North and South thus:

> Slow mortality decline . . . was the result of an accumulation of knowledge, especially medical knowledge, which helped bring infectious diseases under control . . . beginning in the mid-twentieth century, that knowledge slowly accumulated by the rich countries was

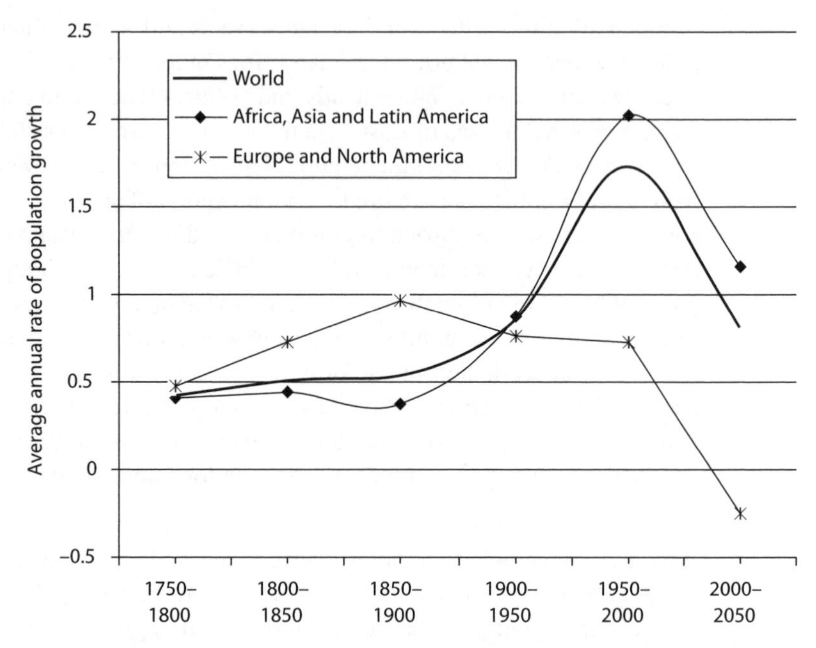

Figure 2.6 Regional trends in population growth rates, 1750–2050

Source: United Nations (1999)

> rapidly transferred to the poor ones and mortality dropped dramatically.
>
> (Livi-Bacci 2001: 129–130)

The rapid decline in mortality in the South beginning in the late colonial period was not immediately accompanied by a decline in fertility rates, which explains the obvious population boom apparent in the graph. As levels of urbanisation rise in low- and middle-income countries, fertility rates will continue to fall, eventually completing the global demographic transition. But at the moment, many low- and middle-income countries are suffering the pangs of the urban transition without the potential benefits of extensive industrialisation.

Figures 2.7 to 2.10 illustrate the progress of urbanisation across regions over time. We see that Latin America was the first former colonial region to experience the shift from a largely rural, agrarian way of life to a predominately urban one. Indeed, levels of urbanisation in the Latin America and Caribbean (LAC) region are now in many cases above those found in Europe (see Figure 2.11). Latin American countries achieved independence from European powers much earlier than African and Asian countries and the independent governments of Latin America employed a

Figure 2.7 World urbanisation, 1890

Figure 2.8 World urbanisation, 1950

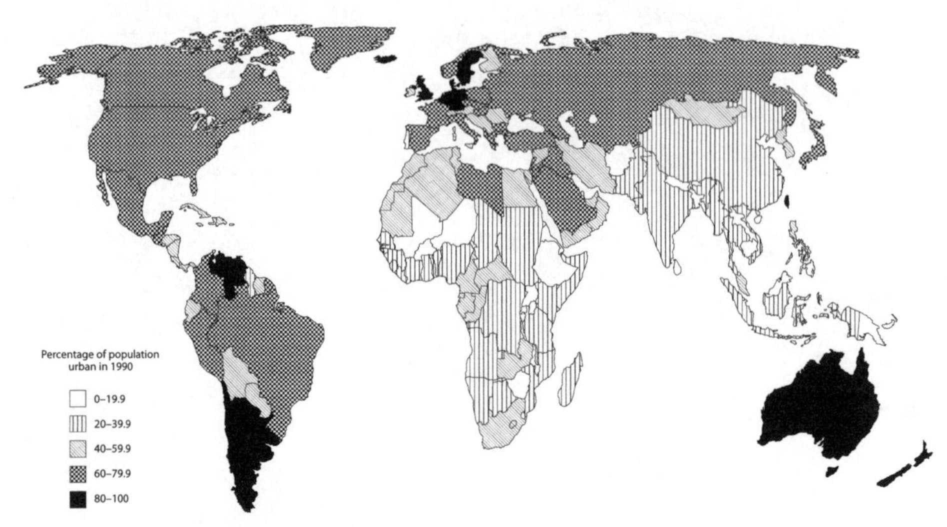

Figure 2.9 World urbanisation, 1990

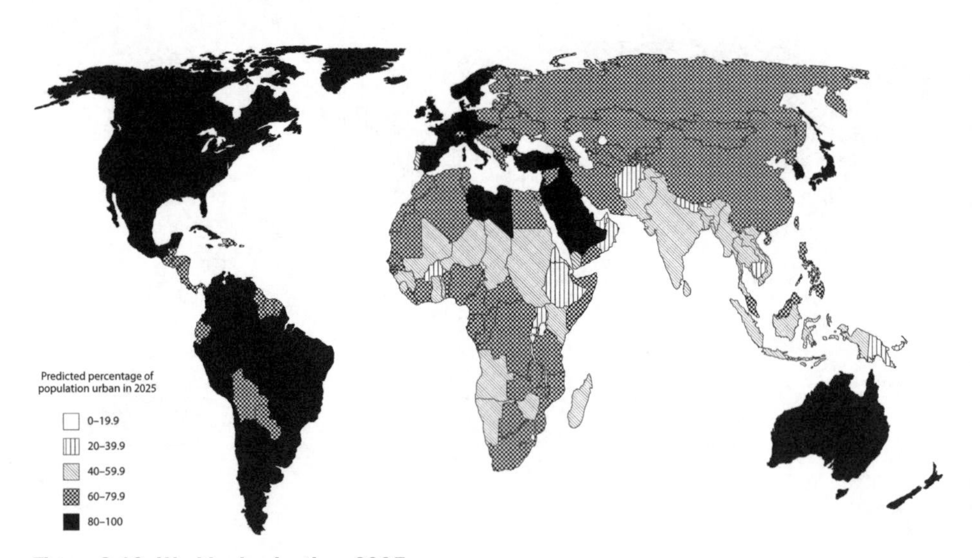

Figure 2.10 World urbanisation, 2005

Source: Pacione (2005)

variety of strategies to stimulate an industrial transformation (with varying degree of success), which also stimulated an urban transition in the nineteenth and early twentieth centuries. Despite their early start, however, many countries in the LAC region continue to struggle with

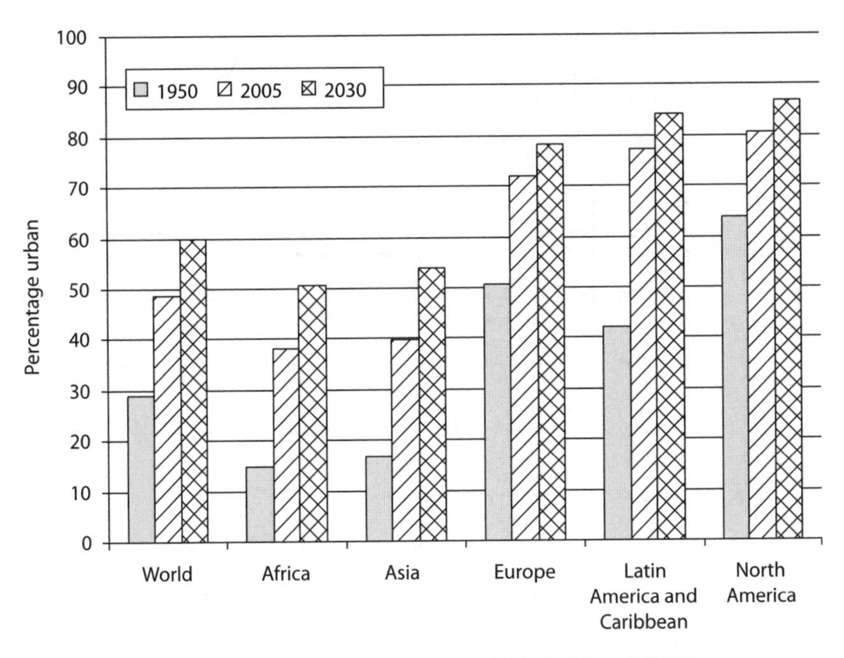

Figure 2.11 Levels of urbanisation by region, 1950, 2005 and 2030

Source: United Nations (2008)

widespread poverty in their cities directly related to the complex legacies of colonialism, rapid urbanisation and urban growth in decades past.

We also see from these figures that Asian and Africa countries remained predominately rural throughout the twentieth century. But by 2030, both regions will pass the tipping point and become predominately urban. As we observed in Chapter 1, there are two phenomena associated with this transition that present serious challenges for governments and development agencies committed to improving welfare and encouraging equitable growth. In low- and middle-income countries slums are expanding rapidly (see Figure 2.12). Without significant intervention, UN-Habitat estimates that the global slum population will top 1.3 billion by 2020. And cities are growing to unprecedented size – the number of cities with 10 million or more inhabitants has risen from two in 1950 to twenty in 2005 – raising the question of whether there is an upper bound on urban growth, and if not, how to effectively manage mega-cities. Concern for the future of urban Africans is particularly acute, where poor economic performance and persistent urbanisation reflect a decoupling of the historical relationship between urbanisation and

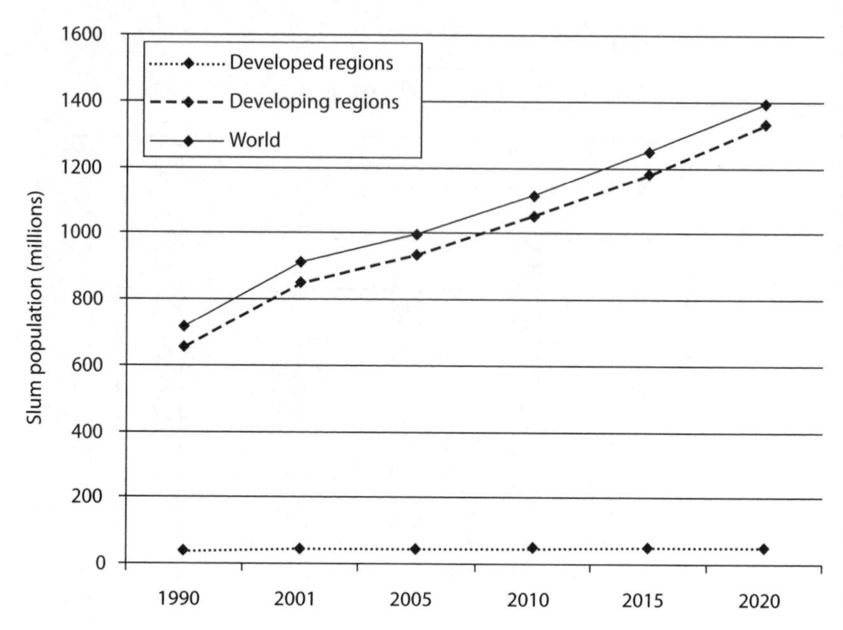

Figure 2.12 Slum population trends and projections

Source: UN-Habitat (2006)

industrialisation – the consequences of which we will explore in greater detail in Chapters 3 and 4.

While mega-cities in low- and middle-income countries seem emblematic of the challenge ahead, they are home to only 9 per cent of the world's urban population and they appear to be growing more slowly than small- and medium-sized cities (United Nations 2008). Indeed, many nations that are still considered predominately rural would be considered more urbanised if the definition of an urban area were adjusted. Urban historians such as Bairoch (1988) often use a lower bound of 5000 inhabitants to classify a settlement as urban. If Pakistan were to adopt this threshold its urban population would jump from one-third to one-half of its total population (Satterthwaite 2006a). The question of whether settlements of this size are large villages or small cities reminds us that strict rural–urban classifications can be problematic. But rapid growth in small settlements is of equal concern, as they are often important regional hubs of political and economic activity and social interaction (Satterthwaite 2006a).

At first glance, then, the picture seems bleak. There are an estimated 11 million people living in Lagos, Nigeria – more than 8 million of them do not have access to piped water in their homes. There are more than 15 million people living in Delhi, India – over 8 million of them do not have access to basic sanitation in their homes. There are some 20 million people living in São Paulo, Brazil – more than 15 million of them do not have ready access to safe water (UN-Habitat 2003a). By 2030, Asia will be home to more than 2.6 billion urban dwellers; Africa will have over 700 million urban residents. Compare this to Europe and North America, which will be home to just over 540 million and 340 million urban inhabitants respectively. The picture is clear: the first urban century will be dominated by cities in low- and middle-income countries and the challenges posed by mega-cities and sprawling slums. We will explore the myriad consequences of this urban landscape in the chapters to follow. But history harbours some hope and offers lessons that we can draw upon to inform our understanding of the contemporary context.

Conclusion

Throughout history cities have been innovative spaces. Urban habitation is defined both by specialisation of social, economic and political activity, and by the nearness of diverse people. The result is change – for better or worse. There is no doubt that early urban societies were highly unequal, and that living conditions were bad. But in the long run, urban innovations such as metallurgy, coined money, writing and bureaucratic administration contributed to the rise of increasingly complex and wealthy societies. From their very origins, cities have been focal points for the accumulation and exchange of information, knowledge, capital and goods. One could make the case that modern society represents the sum total of urban innovations and exchanges across the millennia.

As instruments of empire and colonisation, cities served as command and control centres for the global ambitions of expansive powers from Babylon to Rome to London. In some cases, the residues of segregation and exploitation associated with these episodes in history remain visible in the layout of urban areas, the nature of urban hierarchies, and in the shape of national infrastructure. But exploitation and repression are difficult to maintain in cities unchallenged: the urban-based independence movements in Africa and the Glasgow Rent Strike both demonstrate the urban potential for political mobilisation and transformation. Indeed, the urban transition set in motion by the demographic transition and Industrial Revolution generally have been associated with increasing wealth, welfare and democratisation worldwide – albeit unevenly distributed.

As the twenty-first century progresses, the demographic and economic forces that drive urbanisation and shape urban spaces will continue to be conditioned by global forces set in motion in the sixteenth century. To fully understand and appreciate the complex dynamics shaping urban spaces, societies and economies today, an appreciation of the historical continuities and discontinuities that link cities across the globe is imperative.

Summary

- **The earliest cities evolved around six thousand years ago in Mesopotamia. There is some disagreement as to whether urban centres were a response to the innovations of agriculture or whether they predated and provided the impetus for those innovations.**
- **These cities and those that subsequently emerged in other regions were essentially 'proto-states', sites of political and economic innovation and centres of regional domination and influence.**
- **In medieval Europe the relationship between urban-based capital and rural-based coercion culminated in the rise of the nation-state, as monarchs struck bargains with wealthy urbanites to finance wars and expansion.**
- **The urban transition in Europe was set in motion by both the Industrial Revolution and demographic transition, but subsequently acted as a catalyst for both.**
- **Cities in many low- and middle-income countries evolved somewhat differently due to their role in colonial processes of extraction and domination.**
- **Colonial urbanism led to the growth of ports and capitals, as well as increasing use of urban planning, segregation and the inequitable development of urban infrastructures. The legacy of these practices endures to the present day.**
- **In many low- and middle-income countries the urban transition is occurring without the benefits of the economic advances that accompanied it in Europe. One consequence of this is the proliferation of slums and increases in urban poverty.**

Discussion questions

1 Discuss the relationship between cities and the emergence of agriculture.
2 What role did cities play in the evolution of states in medieval Europe?
3 In the nineteenth and twentieth centuries in Europe, what was the relationship between (a) urbanisation, (b) the demographic transition and (c) industrial development? Are these relationships different in rapidly urbanising countries today?
4 What are the most significant legacies of colonial urbanism for development in contemporary low- and middle-income countries?
5 What can be learnt from the historical experience of urbanisation to inform the future development of cities in low- and middle-income countries in the first urban century?

Further reading

Bairoch, Paul (1988) *Cities and Economic Development From the Dawn of History to the Present*, Chicago, IL: University of Chicago Press.
Childe, V. Gordon (1950) 'The Urban Revolution', *Town Planning Review*, 21(1): 3–17.
Graham, Stephen and Simon Marvin (2001) *Splintering Urbanism: Networked Infrastructures, Technological Mobilities and the Urban Condition*, London: Routledge.
Hall, Peter (1998) *Cities in Civilization*, London: Weidenfeld & Nicolson.
King, Anthony D. (1990) *Urbanism, Colonialism and the World-Economy*, London: Routledge.
Livi-Bacci, Massimo (2001) *A Concise History of World Population*, 3rd edn, Oxford: Blackwell.
Mumford, Lewis (1961) *The City in History*, London: Martin Secker & Warburg.
Tilly, Charles (1994) 'Entanglements of European cities and states', in *Cities and the Rise of States in Europe: A.D. 1000 to 1800*, Boulder, CO: Westview Press.

Useful websites

Historic Cities website contains collections of maps and documents related to the past, present and future of historic cities: http://historic-cities.huji.ac.il/historic_cities.html

International Planning History Society website includes links to the 'Codes Project': www.planninghistory.org

Open History is an interactive website that illustrates the rise and decline of cities worldwide over the past 4000 years: www.openhistory.net

Urban History Association: http://uha.udayton.edu/index.htm

3 Urbanism and economic development

Introduction

Understanding the process of economic development remains a central theme in the field of development studies and is a core concern for policy makers and development professionals. Improvements in material well-being enhance freedom by giving people greater choice over what they do, where they go and what they consume. More wealthy societies also have greater resources at their disposal to finance public goods such as education and healthcare, or to effectively manage shocks such as natural disasters. Although there are many scholars and professionals who feel that the international development agenda often places too much emphasis on economic growth, few would argue with Thomas' assertion that 'there are no examples of large-scale improvements in living standards without industrialisation and the huge dislocations it brings' (Thomas 2000a: 21).

In this chapter we explore the place of cities in the process of economic development, as well as how cities have been understood in relation to regional and global economic processes. In Chapter 1 we observed that the process of urbanisation has generally been understood as a by-product of industrialisation. However, in Chapter 2 we suggested that cities could play a catalytic role in processes of economic change. In this chapter we begin with a more in-depth exploration of the relationship between the processes of urbanisation and economic development before turning to the question of whether or not cities can serve as engines of growth and development. The density and diversity of economic activity in cities generate 'external economies of scale' which can enhance productivity and accelerate innovation. However, urban economies do not function in isolation, but rather interact with rural areas and other cities across the globe. Cities serve as critical nodes in regional and global economic systems, producing, consuming and trading the world's capital, goods, people and ideas. 'City-regions' are increasingly seen as important units of analysis and intervention by those concerned with stimulating economic development and sustained economic growth. But in the wake of neo-liberal reforms, and spurred on by global economic restructuring, there appear to be a growing trend towards the 'informalisation' of urban economies, which raises important challenges in terms of harnessing the economic potential of cities.

Urbanisation and economic development

Historians and economists have long observed a close relationship between levels of urbanisation and levels of economic development (see Figure 3.1). Generally speaking, economic growth and development are regarded as primary stimuli for urban growth and urbanisation. Indeed, for modernisation theorists, urbanisation was understood

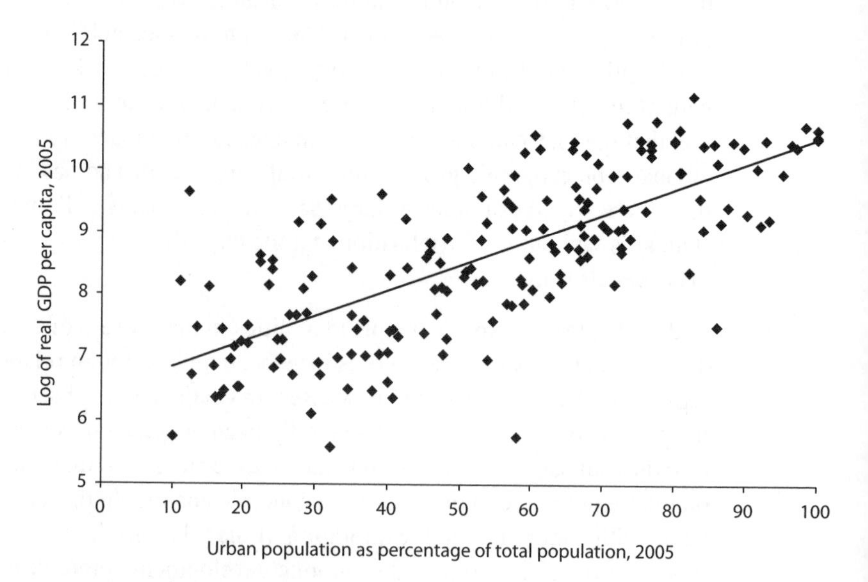

Figure 3.1 Urbanisation and income, 2005

Source: World Bank (2006)

primarily as a by-product of economic development. In order to explain this relationship, we begin with a brief explanation of economic development as a process.

The terms 'economic growth' and 'economic development' are sometimes used interchangeably. It is important, however, to distinguish between the two. Economic growth is defined simply as an expansion of output for a given economy. Economic development entails '*growth plus change*: there are essential qualitative dimensions in the development process that extend beyond the growth or expansion of an economy through a simple widening process' (Meier 1995, emphasis in original). In particular, economic development involves a shift from subsistence agriculture to higher-value-added sectors, such as manufacturing, industry and business services, which results in a more diverse, complex economy. It is important to distinguish between the two because growth can occur without economic development.

For example, a country that produces oil may benefit from rising prices in international markets. If the volume of oil being produced remains constant but the price of oil rises, the value of the country's exports will rise and generate growth. However, unless the profits from such a windfall are reinvested in social development (i.e. health and education) or in other sectors of the economy (thereby encouraging diversification) growth may not stimulate development. Countries where growth is driven by a single commodity – or small number of commodities – are vulnerable to shocks that can undermine their economic base (such as a dramatic fall in the price of the good they produce or a natural disaster). More diverse economies are less vulnerable to such shocks, and are more likely to deliver consistent growth over the long run.

What elements are required to initiate and sustain the expansion of the productive capacity and diversity of an economy? Although it is beyond the scope of this book to provide a comprehensive overview of theories of economic growth and development, some important concepts warrant mention. Early development economists focused first on the role of investment. Expanding production requires investing in direct inputs (such as raw material, tools, buildings, machines) as well as public goods (such as infrastructure and education). It was argued that some countries grew faster than others due to higher levels of savings and investment, which allowed them to accumulate productive resources (Todaro 2000). However, it was subsequently observed that there is a limit to how far investment can sustain growth. Investment appeared to be a necessary but insufficient condition for development. This led to the insight that innovation plays a critical role in the process of economic change (Todaro 2000). New ideas and technologies can improve productivity and generate new kinds of goods and services, allowing an economy to continue expanding when all of the potential economic gains to old ideas and old technologies have been realised. For low- and middle-income countries, the adoption of innovations developed in more advanced economies can substitute temporarily for domestic innovation as these countries 'catch up' with the global technological frontier. But both of these explanations, although theoretically sound, cannot explain the dramatic divergence of economic performance across countries over long periods of time. Why do some countries fail to save, invest and innovate?

To answer this question the current generation of development economists has increasingly focused on the institutional underpinnings of economic development. There is a growing consensus among development scholars that institutions – including the maintenance of law and order, the enforcement of property rights and even social and cultural norms – play a crucial role in sustaining growth and development in the long run (North 1990; Rodrik 2003; Rodrik *et al.* 2004). Generally speaking, institutions provide a framework for productive interactions between economic agents by reducing the uncertainty and costs of transacting with strangers. Institutions also provide incentives for individuals to invest in growth-enhancing activities. For example, individuals who own the land they live on are likely to invest in improving it over time – perhaps by building a house or planting crops. In the pharmaceutical industry, patents (or 'intellectual property rights') give firms incentives to invest in developing new kinds of medicine. In other words, institutions play an important role in defining the quality of the environment for investment and innovation.

Together, investment and innovation (or the adoption of unexploited technologies from abroad) underpinned by an institutional regime that encourages these activities and provides a secure exchange environment facilitate economic growth and development. The resultant process of economic change has important implications for urban change. First, the growth and diversification of an economy entails the production of increasingly complex goods that require a wider range of skills and inputs to produce. This encourages specialisation and exchange, which in turn encourages the concentration of production in a central place – a town or city. As a result, demand for labour in urban areas rises, thereby encouraging urban migration. Second, as an economy grows and people get richer, they begin to consume new kinds of goods. Individuals spend less and less of their incomes (in percentage terms) on food and more on manufactured goods and services, which are produced primarily in cities. Rising demand for non-agricultural goods further encourages the expansion of industry and services. The result is a structural change in the economy: employment in agriculture declines and employment in manufacturing, industry and services rises. This economic transformation – i.e. industrialisation – is invariably associated with urbanisation and urban growth due to the significant economic benefits of agglomeration (discussed in the next section).

In Chapter 2 we noted that the processes of industrialisation and urbanisation ran parallel in Europe. Economic transformation and expansion (alongside demographic forces) stimulated migration to cities as new industries created new jobs en masse. This historical evidence of the relationship between economic growth and urbanisation is further supported by cross-sectional analysis. As demonstrated in Figure 3.1, cross-country comparison shows that at any given point in time, a nation's level of urbanisation is highly correlated with its level of income. Furthermore, if we compare the structure of employment across regions we see that in the poorest and least urbanised regions of the world – Sub-Saharan Africa and South Asia – over 50 per cent of the workforce is engaged in agriculture, while in the richest regions of the world over 95 per cent of the workforce is engaged in industry and services (see Table 3.1). These figures lend support to the notion that cities are by-products of the industrialisation process (Henderson 2003).

But is this conventional wisdom, based primarily on the European experience, the whole story? Many authors have noted that there appears to be no correlation between levels of urbanisation and subsequent rates of economic growth (see Figure 3.2), supporting the theory that urbanisation does not necessarily stimulate economic growth but is evidence of economic progress. And yet there is no question that the urban share of a nation's output (in per capita terms) generally exceeds the rural share. In other words, urban economies are almost always more productive than rural ones, suggesting that urban economies can in fact be 'engines of growth'. For example, São Paulo is home to just 9 per cent of Brazil's population but produces over 36 per cent of the nation's gross domestic product (GDP); Nairobi in Kenya holds just 5 per cent of the nation's population and yet produces one-fifth of the nation's wealth; and while only about 2 per cent of China's population lives in Shanghai, the city produces over 12 per cent of the country's output (Friere and Polèse 2003: 6).

Table 3.1 ***Employment by sector as a percentage of total employment***

	1996			2006		
	Agriculture	*Industry*	*Services*	*Agriculture*	*Industry*	*Services*
Developed economies and EU	5.2	28.5	66.4	3.2	24.2	72.7
East Asia	54.0	25.2	20.7	48.3	25.8	25.8
South East Asia and Pacific	51.0	16.4	32.7	47.0	17.8	35.2
South Asia	59.3	15.4	25.3	51.7	18.8	29.5
Latin America and Caribbean	23.2	20.3	56.5	18.8	19.8	61.4
Sub-Saharan Africa	68.1	9.0	22.9	63.0	8.8	28.2

Source: ILO (2007)

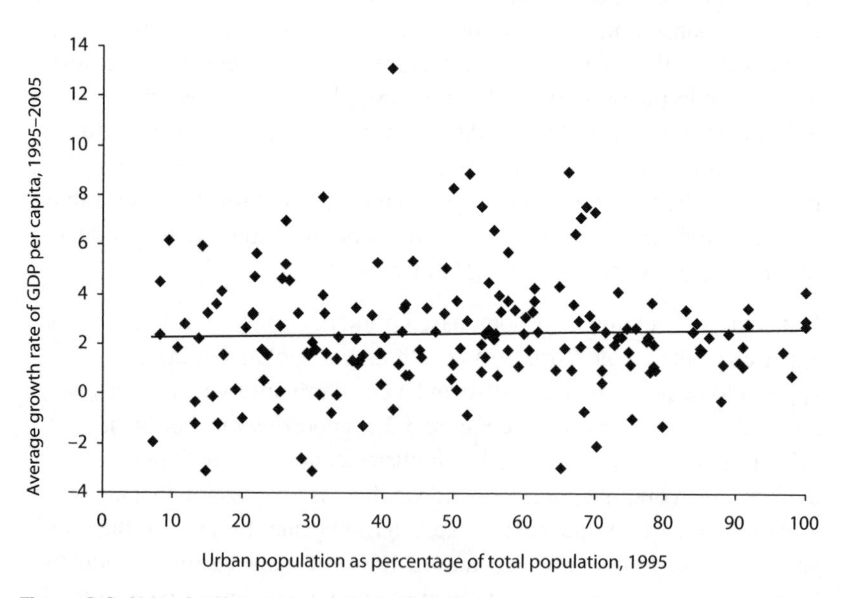

Figure 3.2 Urbanisation and economic growth

Source: World Bank (2006)

Whether or not cities are catalysts for economic growth and development or merely a consequence of economic growth is of particular interest when we consider urbanisation trends in Africa today. Since the early 1980s Africa has been urbanising rapidly despite very poor growth performance. In other words, it appears that non-economic forces are driving urbanisation, and as a result, there appear to be more urban dwellers than jobs or houses. Is Africa 'over-urbanised'?

The concept of over-urbanisation is predicated on an inappropriate comparison between the urbanisation trajectories of countries in the North and those in low- and middle-income countries. An implicit assumption is made that all countries should follow identical paths of urbanisation and industrialisation. The fact that the demographic, industrial and urban transitions happened simultaneously in Europe should not lead us to believe that these transitions always occur simultaneously. And empirically speaking, they do not.

In an attempt to explain Africa's apparent over-urbanisation, Fay and Opal (2000) found that a wide range of factors influence the pace of urbanisation, including rural–urban wage differences (as predicted by Harris and Todaro's (1970) model of migration), the economic structure of a country, levels of education, ethnic tensions, and civil disturbances. And once cities reach a certain size, it appears that they continue to grow, carried on by their own momentum. In other words, African cities are not 'over-urbanised' in any objective sense, but rather have urbanised in response to a range of both economic and non-economic factors. Nevertheless, over-urbanisation remains a useful concept if it is understood not as a deviation from normalcy, but rather a description of the challenges associated with large urban populations and anaemic urban economies.

Just as over-urbanisation has raised concerns, so too has urban concentration, or 'the degree to which urban resources are concentrated in one or two large cities, as opposed to spread over many cities' (Davis and Henderson 2003: 98). The phenomenon of 'excessive concentration' is generally referred to as 'urban primacy'. Primate cities are those that dominate their urban systems in terms of population size, as well as political and economic power. Cities such as Buenos Aires, Lima, Mexico City, Dhaka, Bangkok, Jakarta and Kinshasa are all primate cities in their respective countries. On average Africa, East Asia, Latin America and the Middle East have the highest rates of urban primacy in the world. Concerns about urban primacy have been raised on both equity and efficiency grounds. In terms of equity, an unbalanced urban system raises the spectre of geographically uneven development, leading some planners to argue that settlement patterns should 'be adjusted so as ultimately to provide a regional structure that meets the demand for equality of status for people in different parts of the country' (Hägerstrand 1978: 556). From an efficiency standpoint, economists have explored the possibility that excessive urban concentration can have a negative impact on a nation's economic productivity. Primate cities are, by definition, significantly larger than all other cities in a given country, and their size creates challenges for planners and policy makers trying to ensure the efficient and equitable functioning of such settlements, such as congestion and rising costs of urban land (Henderson 2003).

There is no general consensus regarding the causes of urban primacy. Is it a transitional phenomenon, as suggested by Hirschman (1958)? Some research has shown that urban concentration does indeed tend to change as a country's economy expands and develops. The early stages of urbanisation and economic development are generally characterised by high degrees of concentration, while at higher levels of urbanisation and economic development concentration tends to decline. From an economic point of view, this arc represents an efficient allocation of resources over time. At lower levels of development it makes economic sense for individuals and firms to cluster together to take advantage of limited physical infrastructure and skilled labour, as well as information spill-overs. As an economy (and primate city) grows and infrastructure improves, the prospect of spreading resources more evenly across urban centres becomes both plausible and appealing, as the primate city becomes increasingly expensive to operate in (Henderson 2003). However, Carol Smith (1995: 38) suggests that primacy is not a transitional but rather a historical phenomenon – a legacy of the way in which national territories were integrated into the world economy and the way in which states mediated this process. Studies that have shown that urban concentration tends to be significantly higher in non-democratic countries than in democratic ones lend support to this general thesis. It appears that the concentration of absolute political power in the hands of a few located in a capital city is mirrored by a concentration of people in that city, perhaps looking to influence the leadership or take advantage of the concentration and redistribution of wealth in the capital (Ades and Glaeser 1995; Davis and Henderson 2003).

In response to perceived over-urbanisation and excessive urban concentration, governments have employed a range of policies designed to discourage rural–urban migration and redirect migration away from core cities. These have included legal restrictions on residence in cities, relocations schemes, rural development initiatives and growth pole strategies. History has demonstrated that trying to curtail urbanisation and control urban concentration is rarely possible – with only a few notable exceptions such as China and South Africa, where draconian migration policies were implemented with force. Moreover, there are theoretical and empirical reasons to believe that such efforts may in fact be counterproductive – economists since Alfred Marshall (1920) have noted the positive benefits of urbanism. Instead of working against urban growth, understanding the pros and cons of urbanism can help to inform constructive policies and interventions to ensure that cities work as efficiently and equitably as is possible.

Cities as engines of growth

Jane Jacobs (1984) argued vigorously that cities, not nations, are the salient unit of analysis for understanding economic processes. As she observed:

> Nations are political and military entities, and so are blocs of nations. But it doesn't necessarily follow from this that they are also the basic, salient entities of economic life or that they are particularly useful for probing the mysteries of economic structure, the reasons for rise and decline of wealth.
>
> (Jacobs 1984: 31)

Jacobs (1984: 32) goes on to note that 'most nations are composed of collections or grab bags of very different economies, rich regions and poor ones within the same nation'. In order to understand *inter*-national and *intra*-national disparities of economic development, Jacobs suggests that we examine the economies of cities, which 'are unique in their abilities to shape and reshape the economies of other settlements, including those far removed from them geographically' (1984: 32).

What is unique about the economies of cities? Agglomeration – or the clustering of people and economic activity in space – facilitates the circulation of goods, people, knowledge and ideas. This, in turn, generates what economists refer to as *external economies of scale*, which enhance efficiency and stimulate innovation. External economies of scale refer to the productivity gains realised by a producer or firm due to factors external to a particular producer or firm. An individual firm can improve its own productivity through investment – by buying new technology to improve the efficiency of the production process, for example. But productivity gains can also be a result of factors that an individual firm has no immediate control over. Individual firms do not control the size, density and diversity of a city, and yet these factors improve individual firm performance by generating external economies of scale (see Box 3.1). Economists generally distinguish between two kinds of external economies of scale that are a direct result of urbanism: *localisation economies* and *urbanisation* or '*Jacobs economies*'.

The spatial clustering of firms engaged in the production of a similar good produces localisation economies. The concept of localisation economies grew out of economist Alfred Marshall's (1920) observation that (a) proximity allows firms to share inputs; (b) a large pool of labour improves matching between the needs of firms and the skills of workers; and (c) in densely populated areas people share information and ideas,

Box 3.1

Key definitions in urban economics

Economies of scale refer to the productivity gains realised when making more of a good (i.e. producing it on a larger scale) results in lower unit costs.

Externalities are external benefits or costs that accrue to others as a result of an individual's (or firm's) activity that are not reflected in market prices. Although not strictly accurate, externalities can be thought of as 'unintended consequences'.

External economies of scale refer to the productivity gains realised by a producer or firm due to factors external to a particular producer or firm. The clustering of economic activity in space generates external economies of scale.

Localisation economies are a subclass of external economies of scale. Localisation economies result when the producers of a particular good or service are clustered in space, allowing them to share inputs, draw on a larger pool of labour and benefit from the rapid diffusion of knowledge and innovation that occurs in densely populated areas.

Urbanisation economies are also a subclass of external economies of scale. In this case, however, the benefits derive from the diversity of economic activities concentrated in space, which facilitate the cross-fertilisation of knowledge and ideas thereby stimulating innovation. Innovation, in turn, can lead to greater efficiency for individual producers and is considered a key driver of economic growth more generally.

facilitating learning, innovation and diffusion (Marshall 1920: 267–277). All of these factors – a direct consequence of the close proximity of economic agents – can improve the performance of a firm or industry.

Consider an automobile manufacturer. Few auto companies produce every component that goes into making a car. Instead, they buy many of the individual parts and assemble them in a factory. In a city with just one car manufacturer, a firm that makes wheels would have only one local customer. But if several car manufacturers locate in a city, the wheel manufacturer has a larger market to serve, and hence a stronger incentive to produce more wheels. By making more wheels, the producer can take advantage of 'economies of scale' – a term economists use to refer to situations in which producing more of a good reduces the unit cost – and hence raises productivity. Moreover, in a city with several car manufacturers there is an incentive for more wheel manufacturers to locate there. More manufacturers increases competition (which drives down prices) and creates a reliable source of inputs for the automotive manufacturers, providing further incentives for firms to locate in that particular city. This sharing of inputs is one of the positive feedback mechanisms that create 'industry towns' such as Buffalo City, the site of South Africa's automotive industry, or Bangalore, known as 'India's Silicon Valley', being the country's high-tech industry centre.

The second mechanism that improves firm performance is a better matching of firm needs with skilled workers. If an automotive manufacturer were to locate in a sparsely populated rural area, they would have a limited pool of labour to draw upon – they would take whatever labour they could get. But by locating in a densely populated urban area, a firm can draw from a larger talent pool and can therefore afford to be pickier about whom they hire. The likely result is a more productive staff that will improve firm performance. Again, the productivity gain in this case is a side effect of urbanism. And again there is a positive feedback mechanism in that the concentration of an industry in a particular city encourages labourers to acquire relevant skills, or encourages skilled migrants to gather in the city. A contemporary example of this is Bangalore, India, which has become a world-class centre for the production of information technology (IT) services. Local university students have a strong incentive to acquire computer skills in order to improve their employment prospects, and already skilled workers flock from all over India to the region, encouraged by the many potential employers.

Not only has Bangalore become a world-class producer of IT services but also it has increasingly become a hotbed of innovation, highlighting Alfred Marshall's third point. As he observed, in cities:

> The mysteries of the trade become no mysteries; but are as it were in the air, and children learn many of them unconsciously. Good work is rightly appreciated, inventions and improvements in machinery, in processes and the general organization of the business have their merits promptly discussed: if one man starts a new idea it is taken up by others and combined with suggestions of their own; and thus it becomes the source of further new ideas.
>
> (Marshall 1920: 271)

Proximity, in this case, has the benefit of facilitating the flow of information and ideas, producing yet another form of positive feedback that boosts productivity. The diffusion of knowledge and ideas, as well as innovation, drives economic development.

There are several other ways that an urban context generates localisation economies beyond those originally identified by Marshall (1920). By locating in densely populated areas, firms benefit from being close to the consumers of their final goods, which reduces transaction costs. Transporting goods always carries a significant cost, so the closer a firm is to its consumers, the lower the market price of their goods will be. There are also potential external economies of scale in public investment and the provision of public goods such as infrastructure. Education, healthcare and electricity can be more cheaply provided in urban areas (Arnott and Gersovitz 1986), as can potable water, roads and sanitation systems. Hence, an urban bias in public investment (see Chapter 1) may actually be justified by a simple cost-benefit analysis, as more people are reached in cities than in the countryside for the same level of investment.

All of the above external economies of scale relate directly to the effects of people and firms clustered together in space. Just as density was identified by Wirth (1938) as a defining characteristic of urbanism, so too was heterogeneity (or diversity). The external economies of scale produced by diversity are known as urbanisation economies or Jacobs economies. Jane Jacobs was a great believer in the positive effects of urban diversity and its role in stimulating innovation, arguing that 'economic life develops by grace of innovating' (1984: 39) and that 'development is a process of continually improvising in a context that makes injecting improvisations into every day life feasible. Cities . . . create that context. Nothing else does' (1984: 154). In particular, Jacobs (like Marshall) argued that cities encourage the recombination of old ideas into new ones, thereby stimulating innovation and economic growth.

The important difference between localisation economies and urbanisation economies is that localisation economies can take place in a city with only one industry. Urbanisation economies are a direct consequence of the diversity of firms and individuals in a city, facilitating the cross-fertilisation of knowledge and ideas, generating new ideas, new products and new industries. While it is extraordinarily difficult to accurately measure and assess the effects of diversity empirically, several studies provide robust evidence that diversity does encourages growth at both the industry and city level (see Glaeser *et al.* 1992; Henderson *et al.* 1995; Rosenthal and Strange 2003). Duranton and Puga (2001: 1455) have further demonstrated theoretically and empirically that 'diversified cities act as a "nursery" for firms' by providing an ideal environment for the development of new products, while less diversified cities provide benefits to established firms who seek to avoid the congestion costs associated with large centres.

Globalisation has, of course, significantly reduced the costs of transporting goods, people, information, knowledge and money. Mobile phones, email and the internet have dramatically reduced the costs of and time involved in communication, allowing customers to order goods direct from manufacturers, and making it possible for workers to work from home. Does this portend the death of the city? According to Storper and Venables (2004):

> Face-to-face contact remains central to coordination of the economy, despite the remarkable reductions in transport costs and the astonishing rise in the complexity and variety of information – verbal, visual, and symbolic – which can be communicated near instantly.
>
> (Storper and Venables 2004: 352)

> [It is] a highly efficient technology of communication; a means of overcoming coordination and incentive problems in uncertain environments; a key element of the socialisation that in turn allows people to be candidates for membership of 'in-groups' and to stay in such groups; and a direct source of psychological motivation.
>
> (Storper and Venables 2004: 365)

If face-to-face contact remains essential to the economies of highly urbanised and industrialised nations, as Storper and Venables suggest, it is even more important in less urbanised and less advanced economies in low- and middle-income countries where communication and transportation costs remain high due to limited infrastructure, and insecure institutional environments make long-distance transactions more risky.

This recognition of the economic importance of face-to-face interactions between individuals in an urban context is mirrored by a growing recognition of the importance of relationships between various institutional actors in cities that influence economic productivity. The benefits of agglomeration are not merely derived from the concentration of factors of production in a central place, but also by the cooperative and competitive relationships that emerge between various actors. These spatially contingent relationships can be conceptualised as *clusters*. Porter (2000) defines clusters as

> geographic concentrations of interconnected companies, specialized suppliers, service providers, firms in related industries, and associated institutions (e.g. universities, standards agencies, trade associations) in a particular field that compete but also cooperate.
>
> (Porter 2000: 15)

Porter argues that economic policy should focus on clusters as opposed to sectors, industries or companies because 'a good deal of competitive advantage lies *outside* companies and even outside their industries, residing instead in the locations at which their business units are based' (Porter 2000: 16). He observes that 'connections across firms and industries are fundamental to competition, to productivity and (especially) to the direction and pace of new business formation' and that governments (at local and national levels) have an important role to play in cultivating the 'important linkages, complementarities, and spill-overs in terms of technology, skills, information, marketing, and customer needs that cut across firms and industries' (Porter 2000: 18). In particular, he advocates investment in public goods that benefit all firms in a cluster (e.g. infrastructure and education), maintenance of a business friendly institutional environment and efforts to mobilise and provide support to firms within an identified cluster. Bangalore, India, provides an excellent example of cluster development in the IT sector. Effective cooperation and coordination between urban planners, local government authorities, national government policy, private firms and educational institutes produced a dense web of mutually re-enforcing relationships and incentives that supported the industry's growth (see Box 3.2).

The potential for urban agglomerations to generate external economies of scale, stimulate innovation and cultivate globally competitive economic

Box 3.2

The birth of an IT cluster in Bangalore

Bangalore provides a particularly successful example of cluster development in the IT sector. The city, which is the capital of Karnataka state, was the site of many large-scale investments in manufacturing and industry in the early twentieth century. The infrastructure developed to support these industries, and the presence of skilled engineers in the region, made Bangalore an appealing location for early investments in the growing IT sector beginning in the 1970s. The development of a globally competitive cluster was encouraged by a variety of national and local level policies and initiatives, including financial incentives for firms, targeted education and training to cultivate an appropriately skilled labour force, provision of land and basic infrastructure (including buildings) for potential investors, investments in basic research and development, and promotion of linkages between firms in the sector. Karnataka state was the first in India to develop a comprehensive IT policy, worked cooperatively with city planners and the private sector to develop several industrial parks equipped with the kind of high-tech infrastructure demanded by IT firms, and actively markets Bangalore abroad as an attractive site for foreign investors. Strong links have developed between firms and various educational institutions in the region, who provide appropriate skills training as well as specialised research, some of which has been funded by the local branches of transnational IT firms such as Texas instruments. As a result of these proactive policies, effective coordination and cooperative linkages between firms and educational institutes, Bangalore has become a globally competitive city in the IT sector and a major contributor to the national economy of India. It has, in other words, become a paradigmatic cluster: 'a system of interconnected firms and institutions whose whole is more than the sum of its parts' (Porter 2000: 21).

Sources: Basant and Chandra (2007); Patibandla and Petersen (2002); van Dijk (2003)

clusters is, to some extent, counterbalanced by the costs of agglomeration. Congestion – in the form of traffic jams, for example – represents a significant cost to urban economies, slowing down the flow of goods and people in a city and taking up time that could be used for work or leisure. Estimates from research in the United States suggest that congestion costs result in billions of dollars of economic loses annually. Traffic congestion also contributes to air pollution, which is a classic *negative externality*. Externalities are defined as external benefits or costs that accrue to others as a result of an individual's (or firm's) activity that are not reflected in market prices. For example, in the case of air pollution, an individual's use of automobile transport generates particulate matter that can have negative effects on the health of other urban residents, as well as contribute to the accumulation of greenhouse gases in the atmosphere. But the costs to the general public are not factored into the individual's decision to use a car. In a similar vein, poor water and sanitation infrastructure can generate negative externalities by allowing the growth and transmission of diseases, which reduces labour productivity across urban populations (not to mention creating miserable conditions for urban residents).

Crime rates are also generally higher in urban than in rural areas, and crime rates tend to be higher in larger than in smaller cities. Just as urban density generates external economies of scale for legitimate industries, cities are appealing environments for criminals. A higher density of potential victims and potential customers of stolen or illicit goods, and a lower probability of arrest (due to a higher number of potential suspects), attract the criminally minded (Glaeser and Sacerdote 1999). Insecurity of property and person can, in turn, deter potential investments and divert resources away from productive activities and towards private protection. In South Africa, where surveys of private firms consistently rank crime as one of their primary concerns, a study in 1996 found that there are 900 private security personnel employed for every 100,000 people – more than three times the number found in the UK, Germany or Canada, and approximately twice as many as are found in the United States or Bulgaria (De Waard 1999).

We will address challenges associated with managing these costs in subsequent chapters. The important point, for now, is that cities can be engines of growth and development when governed well.

Rural–urban linkages and regional development

The 'urban bias thesis' discussed in Chapter 1 was fundamentally motivated by a concern with observable disparities in welfare between those living in rural versus urban areas, and the fact that the majority of poor people in the world live in rural areas. Cities were portrayed as parasitic, drawing in and consuming resources produced beyond city limits without adequate reciprocation. This is, however, an overly simplistic portrayal of rural–urban dynamics that overlooks the ambiguities of a strict categorisation of 'rural' versus 'urban' as well as the many positive synergies between rural and urban economies. Where does an urban agglomeration end and a rural settlement begin? There are, of course, administrative boundaries drawn by governments to distinguish between the two, but these do not always accord with reality on the ground. Human settlements are better understood as forming a continuum, from isolated households in far-flung rural areas all the way up to mega-cities. More importantly, however, the distinction becomes increasingly problematic (from an analytical perspective) when we consider the flows of goods, people and resources between rural and urban areas (Lynch 2005; Tacoli 1998).

Cities ultimately depend upon surplus food production in rural areas, but cities are not simply parasites. Historically, urban settlements have helped boost agricultural productivity through the production and distribution of farm equipment and fertilisers (Bairoch 1988), as well as through their role as regional 'service centres' providing access to goods and services (such as banking, medical, repair and information services) that support rural enterprises. Networks of towns and cities 'form an essential marketing network through which agricultural commodities are collected, exchanged, redistributed' (Rondinelli 1994: 373). Producers sell their goods in small market towns where distributors collect and bundle them into larger units, which are sold in larger towns and cities. There they may be consumed, processed or rebundled into yet larger units for sale on international markets. In other words, networks of towns and cities 'are essential to the whole chain of exchange on which commercial agriculture depends' (Rondinelli 1994: 373).

Rural and urban dwellers also rely on each other as markets. The sale of agricultural surplus to urban populations provides extra income to farmers who would otherwise rely on subsistence production. This income may, in turn, be used to purchase not only inputs, but also consumer goods (such as clothing, furniture, cookware, radios, televisions) that are produced in urban areas and travel through the very same urban networks to reach rural consumers. The terms of trade between rural and urban areas have been distorted in the past through government intervention, as argued by Lipton (1977) and Bates (1981) (see Chapter 1) but this does not imply that the economic relations between urban and rural areas *must* necessarily be of an exploitative nature.

There is also a tendency to use the terms 'rural' and 'agricultural' synonymously. In fact, many rural households have diverse livelihood strategies that include producing, trading and marketing a range of non-agricultural goods. Similarly, urban residents – particularly those living in peri-urban areas with access to larger plots of land – often engage in urban agriculture as a component of their livelihood strategies.

Moreover, households may straddle the rural–urban divide, taking advantage of livelihood and entrepreneurial opportunities available in each. In some cases, this may entail the permanent migration of a rural family member to a city, who then sends home remittances from his/her waged employment. In other cases, it may involve seasonal migration, with individuals moving between urban and rural areas at different times of the year to take advantage of waged employment while still participating in agricultural production during labour-intensive seasons, such as harvest time (Tacoli 1998). In yet other cases members of families or larger groupings engage in exchange and trading activities between rural and urban areas.

The potential for dynamic interactions between rural and urban areas to stimulate local/regional development was recognised by urban and regional planners in the early post-war era. From the 1950s to the 1970s, urban and regional planning was seen as way to encourage the integration of domestic economies and facilitate national development more broadly (Friedmann 1967). Investment in 'propulsive industries' and infrastructure in select urban centres was expected not only to benefit urban populations, but to drive regional development more generally by serving as 'growth poles' for wider city-regions (Parr 1999a). Regional planners thought that encouraging decentralised industrialisation in peripheral regions could reduce spatial inequalities. However, growth pole strategies, popular in the 1950s and 1960s, were replaced by the rural development paradigm in the 1970s and 1980s.

Dissatisfaction with industrial decentralisation and growth pole strategies stemmed from inconsistent results in practice. In theory, towns and cities can be regional drivers of change by facilitating agricultural development while providing alternative income-generating opportunities for rural dwellers. However, in practice, growth poles were often set up in depressed areas with low levels of infrastructure and a poor fiscal base. At worst, small towns and cities have been deemed 'vanguards of exploitation' (Southall 1988), serving as gateways for the penetration of the interests of opportunistic global capital into the countryside. More moderate perspectives suggest that the relative success of such initiatives in cultivating regional development largely depends on the local political and economic dynamics of the regions affected. Land ownership structures, national government policies, global economic trends and local social relations may all confound well-intentioned efforts to stimulate local or regional development (Faguet 2003, 2004; Hinderink and Titus 2002; Tacoli 1998). Parr (1999b) attributes the general failure of industrial decentralisation to the uncritical way in which the idea of growth poles was adopted and put into practice. Often planning horizons were too short or plans were overambitious, poorly designed and demanded unrealistic capital outlays. Another handicap was that decisions on the location of growth poles were made on political rather than economic grounds and in many cases the administrative capacity required to see a growth pole strategy through was simply lacking (Parr 1999b).

Despite the mixed record of industrial decentralisation and growth pole strategies, there have recently been calls for a revaluation of urban and regional development strategies. Satterthwaite (2006a: 35) has suggested that we need to 'forget the rural–urban divide and see all settlements as being on a continuum.' Friedmann (2007) argues that we must recognise the 'organic relation' between cities and countryside and makes a plea for policies and interventions that promote the endogenous development of city-regions through the cultivation of their unique assets (see Chapter 8). Scott and Storper (2003: 580) claim that reigning paradigms in development studies have generally overlooked the 'geographical foundations of economic growth' and undervalued the role of cities and regions as 'springboards of the development process'. Active efforts to foster industrialisation and economic development today increasingly take cognisance of the benefits of agglomeration and proximity, seeking to build on areas of economic success or where the potential for spontaneous and endogenous growth is clear. The growing body of theoretical and empirical research demonstrating the economic significance of agglomeration and clustering, a realisation that world urbanisation is an inexorable process, and increasing concerns about the impacts of globalisation on development, all lend a sense of urgency to this agenda.

Cities and regions in a globalising world

Globalisation – or the increasing social, political and economic integration of the world's population – arguably began thousands of years ago when inter-regional trade between settlements in Asia, Europe and the Middle East began. But revolutions in transportation and communication technologies over the past two hundred years, as well as the emergence of new international institutions governing trade and finance in the post-Second World War era, have led to a dramatic acceleration of the process of global integration since the 1950s. Nowadays, global financial markets allow capital to range freely with little regard for national boundaries; advancements in transportation and communications technologies permit an ever-expanding range of goods and services to be traded in unprecedented volumes; transnational and multinational corporations (TNCs and MNCs) have restructured production processes, spreading operations across the globe to exploit the comparative advantage of various regions; workers – from the unskilled to the highly skilled – are increasingly mobile. Urban scholars have taken a particularly keen interest in the processes and effects of globalisation, exploring both the ways in which cities serve as agents of globalisation, as well as the impact that than an increasingly integrated world economy has on individual cities.

As we observed in Chapter 2, cities have functioned as nodes in inter-regional and international trade networks since their first appearance some six thousand years ago. Indeed, as renowned social and economic historian Fernand Braudel (1984: 27) observed, 'A world-economy always has an urban center of gravity, a city, as the logistic heart of its activity'. This idea was further developed by a range of scholars working within the 'world cities' paradigm – a major field of urban research since the 1980s (see Friedmann and Wolf 1982; Friedmann 1986; Hall 1966; Knox and Taylor 1995; Taylor 2004). Eschewing the idea that contemporary globalisation is simply a variation on a much longer playing theme of capitalist development, these scholars maintained that the current era of globalisation represents a 'historically unprecedented' moment in which capitalist institutions have managed to 'free themselves from national constraints' leading to a radical reorganisation of the international space economy (Friedmann and Wolf 1982: 310). In this new global landscape, cities are believed to serve as 'basing points' for the accumulation and deployment of international capital in the hands of TNCs and MNCs, while simultaneously experiencing social and spatial transformations linked to their specific role in the new international division of labour (Friedmann 1986).

This body of research has largely been focused on the ways in which certain major urban centres (primarily in advanced economies) are increasingly becoming command and control centres in the world economy, concentrating the corporate headquarters of global firms in the financial, transport, business service and communications sectors. Much of this research has been preoccupied with developing league tables and mapping the international hierarchy of world cities based on the concentration of global service firms in cities, the interconnectedness (in terms of transport and communications) of cities, and of course the

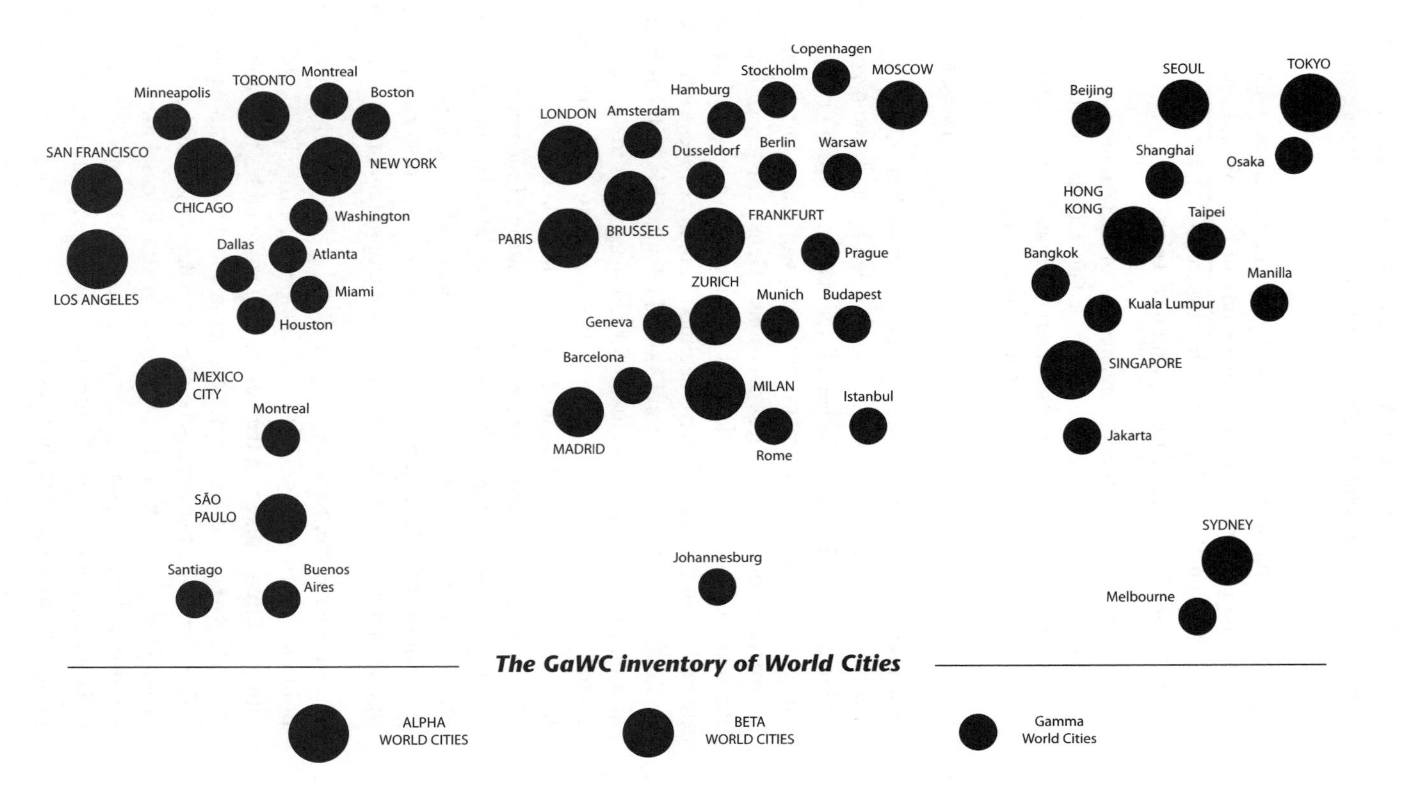

Figure 3.3 The world city hierarchy according to GaWC

Source: Beaverstock *et al.* (1999)

concentration of financial capital in cities (see Figure 3.3 for an example). In these formulations, cities are often ranked according to their apparent role as primary/secondary, core/peripheral or alpha/beta/gamma world cities.

At the top of the world city hierarchy are 'global cities' – a term coined by sociologist Saskia Sassen in her study of New York, London and Tokyo (Sassen 1991). According to Sassen (2006), there are about 40 global cities in the world today, which serve as

> (1) command points in the organization of the world economy; (2) key locations and marketplaces for the leading industries of the current period – finance and specialized service firms; (3) major sites of production, including the production of innovations, for these industries.
>
> (Sassen 2006: 7)

The high concentration of wealth and power in these global and world cities (GaWC) has been accompanied by stark socio-economic polarisation within them. Indeed, Friedmann and Wolff (1982: 322) have claimed that 'the primary fact about world city formation is the polarization of its social class divisions.' Dense clusters of firms in the financial, business service and IT sectors have emerged while manufacturing industries have declined in many of these cities, leading to a decline in demand for semi-skilled labour. This shift has been accompanied by a rising demand for low-wage jobs to service the new global elite and their 'expensive restaurants, luxury housing, luxury hotels, gourmet shops, boutiques, French hand laundries' (Sassen 1991: 9). The result has been a dramatic stratification of incomes in these 'strategic nodes' in the global economy. Social polarisation in cities has always existed, but 'what we see happening today takes place on a higher order of magnitude, and it is engendering massive distortions in the operations of various markets, from housing to labour' (Sassen 2006: 9).

The pervasive fascination with global and world cities has inspired urban politicians across the world to attempt to mimic the success of these cities and find a place on the numerous league tables that rank city status in the global economy, despite concerns about the possible polarisation effect. Far less attention has been paid to the 'large number of cities around the world which do not register on intellectual maps that chart the rise and fall of global and world cities' (Robinson 2002: 531). Cities in less economically advanced nations have generally been regarded as peripheral to the governance and functioning of the global economy. And yet these cities have also undergone significant transformation in the contemporary global era and do play a role in its functioning. As Jennifer Robinson (2006) has argued:

> It is one thing . . . to agree that global links are changing and that power relations, inequalities and poverty shape the quality of those links. It is quite another to suggest that poor cities and countries are irrelevant to the global economy.
>
> (Robinson 2006: 101)

There have been some attempts to balance the world cities/global cities paradigm through an examination of globalising cities in low- and middle-income countries, such as Mexico City, São Paulo, Cairo, Johannesburg, Mumbai, Shanghai and Jakarta (see Gugler 2004; Segbers 2007). These 'second-tier' world cities are playing increasingly important roles in the global economy as centres of production, consumption, regional financial networks and specialised service provision. For example, Shanghai is emerging as a strategic gateway into the rapidly expanding Chinese economy, serving as a regional financial and service centre. At the same time, Shanghai is also experiencing the socio-economic polarisation observed in the global and world cities of more economically advanced nations (see Box 3.3).

Box 3.3

Shanghai: (re-)emerging world city

In 1843, China was forced to open itself up to trade with the West with the Treaty of Nanjing. Shanghai, a small coastal fishing village with a particularly appealing natural port, was dramatically transformed over the following years, eventually coming to be known as the 'Paris of the East'. As an emerging centre of industry and international trade, Shanghai's population mushroomed: by 1900 the city population had reached 1 million people, and by 1936 it was ranked the seventh largest city in the world with a population over 3 million. Over this period Shanghai produced half of China's industrial output, accounted for half of the nation's trade and become the nation's leading financial centre, host to some 90 per cent of China's banks.

In 1949 China took a dramatic turn and closed itself to trade with the West. Shanghai was again transformed. From an international centre of production and trade, the city's economic prowess was directed towards serving the interests and needs of a highly centralised socialist state. As one of China's most productive regions, over 85 per cent of Shanghai's revenue was appropriated by the central state over the subsequent three decades, contributing approximately one-sixth of the central state's revenue over that period.

Beginning in the late 1970s, China began slowly reopening its economy to the world, and again, Shanghai was transformed. Having established a strong industrial base, the city was well placed to take advantage of new opportunities in a globalising economy, becoming a major centre for export manufacturing. In 1990, the central state announced plans to develop a new economic hub in the Shanghai city region, accompanied by a series of reforms handing greater to power to Shanghai's local government. Over the subsequent decade, Shanghai began its re-emergence as a world city. Changes in national laws concerning foreign investment, massive investments in local infrastructure, a wide range of locally promulgated policies to provide incentives for investment, and reforms in land management practices have ushered in a flood of foreign investments. More than half of the world's top 500 transnational corporations and 57 of the largest industrial enterprises in the world have set up shop in Shanghai, contributing to an annual regional growth rate of over 20 per cent – more than twice the national average. In 2007, Shanghai was once again ranked the seventh largest city in the world with a population over 15 million.

The rapid transformation of Shanghai's economy – and skyline – over the course of the 1990s was accompanied by a marked shift in the socio-economic profile of the city. Rising prosperity stimulated migration to the city, attracting over 2 million new migrant workers over the course of the decade. At the same time, the city's manufacturing sector contracted (shedding almost 1 million jobs) while the business services, finance and real estate sectors expanded. With rising demand for high-skilled labour and extensive in-migration of low-skilled labour, the gap between the rich and poor has risen dramatically. Although the city is not considered 'polarised' in the strictest sense, Shanghai's experience does lend support to the general hypothesis that world city formation is inevitably accompanied by socio-economic stratification.

Sources: Li and Wu (2006); Wu (2000)

Shanghai, however, is not merely an isolated pretender to world city status, but rather is part of an emerging 'global city-region' that encompasses many settlements in China and increasingly in neighbouring countries. The concept of a 'global city-region' can be thought of as a corollary to the world or global city concept, but on a broader geographical scale. Global city-regions are 'dense nodes of human labor and communal life' (A. Scott 2001: 1) that manifest as very large metropolitan centres, as well as regional networks of cities (Scott *et al.* 2001: 11).Generally speaking, research on global city-regions is somewhat more inclusive than that related to global and world cities. Using metropolitan size as a rough approximation to measure global city-region status, today there are more than 300 global city-regions worldwide – the overwhelming majority of them in low- and middle-income countries (Scott *et al.* 2001). These city-regions are attracting increasing attention not only because of their importance to the functioning of the global economy, but also because they are increasingly recognised as representing 'distinctive political actors on the world stage' (Scott *et al.* 2001: 11). Since the rise of nation-states in Europe several hundred years ago, states have generally been seen as the most important actor influencing the territorial organisation and regulation of production and exchange. In the contemporary era, some see cities and city-regions as reasserting their historic role in directing the world economy (Segbers 2007: 13).

While some city-regions contain within them the global and world cities that have attracted so much attention, others do not, but nevertheless play important roles in the global economy. For example, the El Paso–Juárez city-region, with a combined metropolitan population of approximately 2.5 million people, consists of two cities sitting just across from each other on the Rio Grande – the river that defines the US–Mexican border. This pair of cities, along with many others scattered along the frontier, play a vital role in managing a large volume of cross-border production and exchange activities (see Box 3.4).

Box 3.4

The economy of the El Paso–Juárez city-region

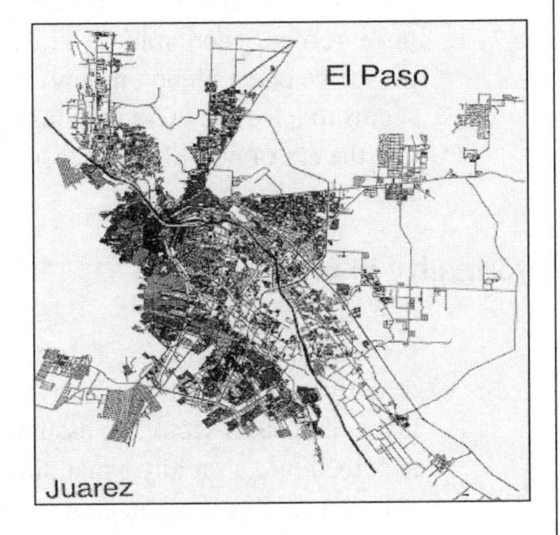

Linked by a handful of bridges that straddle the Rio Grande, the economies of El Paso, Texas and Ciudad Juárez, Mexico are deeply intertwined. This border city-pair is one of many scattered along the 2000-mile frontier that separates the United States from Mexico. Beginning with a Border Industrialization Program launched in 1965, and expanding in the 1980s as the government of Mexico began a process of liberalising trade and reducing restrictions on foreign investment, cross-border trade has stimulated the growth of dynamic border city-regions. In particular, manufacturing plants known as *maquiladoras* sprung up all along the Mexican side of the border, serving as export assembly plants for a wide range of goods destined for US markets including textiles, automobiles and electronics.

The development of the *maquiladora* industry in Mexico has made significant contributions to the economies of both Mexico and the South-Western United States. Employment in *maquiladoras* represents around 30 per cent of all manufacturing employment in Mexico, and the industry produces over $80 billion worth of exports annually. On the US side, the industry has generated demand for finance, accounting and legal services, as well as a host of other services such as hotels, car rental and restaurants. It has also increased demand for US manufactures that are used as inputs in the *maquiladoras*.

The El Paso–Juárez city-region is the largest bi-national metropolitan region in the Western Hemisphere, managing an annual cross-border trade worth over $50 billion – equal to nearly a quarter of all US–Mexico border trade. Everyday more than 2000 trucks, 38,000 cars and 20,000 pedestrians cross the border at El Paso. The growth of the *maquiladora* industry has created over 200,000 jobs in manufacturing in Juárez, and hundreds of thousands of jobs in related industries and services. In turn, El Paso has seen a steady decline in unemployment rates despite a steep decline in manufacturing employment due to the growth of other sectors related to border trade.

Sources: Hanson (2001), Staudt (2001), Vargas (1999, 2001); El Paso Regional Development Corporation. Image: City of El Paso Planning Department

A notable feature of the El Paso–Juárez regional economy is the high incidence of 'informality' – a phenomenon highlighted in Sassen's analyses of global cities and a subject of much research and debate among scholars of urbanism in low- and middle-income countries. Research conducted in the mid-1990s suggests that somewhere between 30 and 40 per cent of the residents of the El Paso–Juárez city-region rely on 'informal' economic activities to get by (Staudt 2001). Informality is a pervasive phenomenon in low- and middle-income countries, and one that seems to have grown in response to the combined forces of globalisation and the era of neo-liberal reform.

Informality in cities

When Arthur Lewis (1954) proposed his two-sector model of economic development, with unproductive rural labourers moving into a dynamic urban-industrial sector, he assumed that the rural migrants would be absorbed into a rapidly expanding urban economy. But since the 1950s, it has appeared as if urban economies were failing to keep pace with urban growth, leading to widespread under-employment and unemployment, and raising concerns about the negative consequences of over-urbanisation. Rather than finding employment in manufacturing and industrial enterprises, most urban dwellers seemed to be engaged in low-productivity activities, eking out a living on the margins of the city economy. Indeed, the following portrait of the vibrant economic life of city streets drawn from an International Labour Organization (ILO) report will be familiar to anyone who has worked, travelled or done research in a low or middle-income city:

> The streets of cities, towns, and villages in most developing countries – and in many developed countries – are lined by barbers, cobblers, garbage collectors, waste recyclers, and vendors of vegetables, fruit, meat, fish, snack-foods, and a myriad of non-perishable items ranging from locks and keys to soaps and detergents, to clothing. In many countries, head-loaders, cart pullers, bicycle peddlers, rickshaw pullers, and camel, bullock, or horse cart drivers jostle to make their way down narrow village lanes or through the maze of cars, trucks, vans, and buses on city streets.
>
> (ILO 2002: 9)

From São Paulo to Nairobi to Cairo to Manila, millions of people rely on these types of activities for their livelihoods. For the most part they operate in the 'informal economy' (also called the shadow economy, second economy or parallel economy). The concept of the informal economy has evolved since the late 1960s, beginning with the introduction of the term 'informal sector' in a 1971 ILO report on employment in Kenya, which was further popularised by anthropologist Keith Hart (1973) in a study of urban employment in Accra, Ghana. In the face of high costs of living and limited formal employment opportunities, Hart found that informal employment was a critical source of income for many urban residents and constituted a significant amount of urban economic activity. The key factor distinguishing formal from informal employment was 'whether or not labour is recruited on a permanent and regular basis for fixed rewards' (Hart 1973: 68) – i.e. self-employment versus waged labour. The ILO report offered a more comprehensive (but perhaps overly rigid) definition of the informal sector, listing the following characteristics:

(a) ease of entry;
(b) reliance on indigenous resources;
(c) family ownership of enterprises;
(d) small scale of operation;
(e) labour-intensive and adapted technology;

(f) skills acquired outside the formal school system; and
(g) unregulated and competitive markets.

Hart (1973), and most researchers since, are careful to note the difference between 'legitimate' informal income opportunities and 'illegitimate' ones. Castells and Portes (1989) distinguish between formal, informal and criminal (or illegal) economic activities. This distinction is important because the overwhelming majority of informal economic activity in cities in low- and middle-income countries revolves around activities that, if recorded and regulated by governments, would be considered wholly legal.

Implicit in the ILO definition above, and in the work of many researchers studying informality, is the notion of a bifurcated or dual urban economy. On the one hand there is assumed to be a modern, capital-intensive, formally regulated, productive tier with relatively secure employment, and on the other a 'traditional,' labour-intensive, unregulated and marginally productive lower tier. Some trace this dualism back to the colonial origins of urbanism in low- and middle-income countries (Drakakis-Smith 2000: 125; Potts 2007), although research has shown that the conceptual distinction is often blurred in practice. Many studies have illustrated the interdependence and close linkages between informal and formal economic activity in industries as diverse as textiles, food service, and electronics (see Portes and Haller 2005), as well as labour mobility between the informal and formal sectors (Maloney 1999). Recognising these complexities, the term informal sector has given way to 'informal economy', which can be defined as

> those economic activities that circumvent the costs and are excluded from the benefits and rights incorporated in the laws and administrative rules covering property relationships, commercial licensing, labor contracts, torts, financial credit and social security systems.
>
> (Feige 1990: 992)

This shift reflects the influence of 'legalist' or neo-liberal perspectives on informality. The relationship between the formal and informal economies is regarded by those on the left as exploitative and likely to perpetuate poverty, while those on the right stress the constructive entrepreneurial initiative of those operating in the informal sector – initiative that is put at risk by state interference in markets. Hernando de Soto's *The Other Path* (1989) was a particularly influential work in this regard. He argued that the informal economy was a product of over-regulation by states. In contrast to the dual economy perspective, which emphasised the failure of the modern sector to intervene to create jobs as generating informality, de Soto's neo-liberal perspective emphasised excessive regulation, bureaucratic inefficiency and corruption as drivers of the informal economy, as people sought to escape the reach of the state.

In recent years the World Bank has been collecting data on how difficult it is for individuals to establish formal enterprises, and the evidence suggests that state regulation can discourage formalisation in some cases. Table 3.2 provides a snapshot of some of the challenges facing an individual who is looking to start a formal business. In Angola, this requires 13 different procedures, 124 days, and almost 500 per cent of the average yearly income of an Angolan. By contrast, in the United States the same process requires just 5 procedures, 5 days and 0.7 per cent of the average annual income of an American. Confronted with high costs to entry, many entrepreneurs may well choose informality. In 2007 a World Bank study of informality in Latin America found that self-employed workers and micro-enterprises *prefer* the informal sector, as it offers more autonomy, flexibility, stability and mobility (Perry *et al.* 2007). The picture is different, however, for waged labourers in the informal sector. The same study found that these workers often aspire to the security of income and conditions of service afforded by formal employment that are simply not available in the informal economy. The study proposes an analytical framework whereby the first group choose to *exit* the formal economy due to the advantages of informality, while the second group suffer *exclusion* (Perry *et al.* 2007).

The establishment of this middle ground is reflected in the definition of informal employment adopted by the ILO, and by recent trends in

Table 3.2 ***Costs of starting a business in selected countries***

	Procedures (number)	*Time (days)*	*Cost (% of income per capita)*
Angola	13	124	486.7
Bolivia	15	50	140.6
Cambodia	10	86	236.4
DR Congo	13	155	481.1
Venezuela	16	141	25.4
Yemen	12	63	228.0
Japan	8	23	7.5
United Kingdom	6	18	0.7
United States	5	5	0.7

Source: World Bank (2005)

development policy and practice. Nowadays the ILO defines informal employment more broadly as 'both self-employment in informal enterprises (i.e., small and/or unregistered) and wage employment in informal jobs (i.e., without secure contracts, worker benefits, or social protection)' (ILO 2002: 7). And while international development agencies such as the World Bank and IMF continue to push for liberalisation, privatisation and deregulation, NGOs and country governments have embraced the idea of supporting the economic everyman through microfinance and small-medium enterprise development schemes to help bolster the opportunities of enterprising individuals. Unfortunately this shift has sometimes been at the expense of attending to fundamental welfare issues – such as decent work conditions, health, education and social protection – that remain largely unresolved. Small-scale entrepreneurs are particularly vulnerable to health crises, rarely have access to any form of pension or retirement fund, and are limited in their ability to move up the value chain without access to education or specialised training. Ad-hoc, small-scale efforts to support entrepreneurs can generate some benefits, but they are unlikely to be sustainable without more comprehensive reforms. Structural adjustment programmes, which were expected by their champions to reduce the size of the informal sector by removing the supposed distortions leading to informality, appear instead to have expanded informal sector activity (Meagher 2003). And just as globalisation has created space for 'world cities' to expand their role in the operations of the global economy, informal producers and traders have taken advantage of increased integration to expand into global markets (MacGaffey and Banzenguissa-Ganga 2000; Meagher 2003). Indeed, informalisation is seen by some as a globally connected process in which formal sector profits are protected by increasing recourse to informal production, with the tacit support of governments in low- and middle-income countries (Meagher 1995).

It is difficult to accurately measure the volume of informal activity in a nation's economy due to the very nature of informality. However, there is a general consensus that informality rose significantly during the neo-liberal reform era of the 1980s and 1990s in many low- and middle-income countries, and that the informal economy represents a significant degree of economic activity in those countries (ILO 2002; Perry *et al.* 2007). For example, the ILO estimates that informal employment represents 72 per cent of all non-agricultural employment (a rough proxy for urban employment) in Africa, 65 per cent in Asia and 51 per cent in Latin America (see Table 3.3).

Table 3.3 ***Informal employment as a percentage of non-agricultural employment***

Region/country	*Informal employment as a percentage of non-agricultural employment*	*Women's informal employment as a percentage of non-agricultural employment*	*Men's informal employment as a percentage of non-agricultural employment*
Sub-Saharan Africa	***72***	***84***	***63***
Benin	93	97	87
Chad	74	95	60
Guinea	72	87	66
Kenya	72	83	59
South Africa	51	58	44
Latin America	***51***	***58***	***48***
Bolivia	63	74	55
Brazil	60	67	55
Chile	36	44	31
Colombia	38	44	34
Costa Rica	44	48	42
El Salvador	57	69	46
Guatemala	56	69	47
Honduras	58	65	74
Mexico	55	55	54
Venezuela	47	47	47
Asia	***65***	***65***	***65***
India	83	86	83
Indonesia	78	77	78
Philippines	72	73	71
Syria	42	35	43
Thailand	51	54	49

Source: ILO (2002)

There has also been increasing recognition and concern regarding informality in markets other than labour. Formal land and housing markets, for example, have failed to keep pace with urban growth as well, resulting in substantial informality in these sectors. All of the statistics regarding the size and growth of slum populations in this book are derived from recent research by UN-Habitat, which initiated the data collection exercise in an attempt to measure the security of tenure of urban dwellers around the world. Due to the enormous costs and complexities associated with trying to develop an index capturing tenure security, it was decided that housing deprivations were the best proxy available. By this measure some 1 billion urban dwellers worldwide are relying on informal land and housing markets to provide a roof over their heads.

Informality in land and housing markets is a product of poorly defined or inefficiently managed property rights regimes. Ambiguity and insecurity of property ownership and occupation can translate into heavy social and economic costs. For example, without an address people find it difficult to obtain work, official documents and even citizenship. Without secure housing people are denied collateral on loans or financial support as entrepreneurs.

It is very difficult to directly measure the costs of informality in land and housing markets, but several studies suggest that they are high (Becker and Morrison 1999). Hernando de Soto (2001) has argued that there are enormous economic resources lying dormant in slums around the world in the form of 'dead capital' – assets whose potential cannot be realised because they are not recognised by a formal property rights system. Where slum dwellers do not have legal rights to their land and property,

> assets cannot readily be turned into capital, cannot be traded outside of narrow local circles where people know and trust one another, cannot be used as collateral for a loan and cannot be used as a share against an investment.
>
> (de Soto 2001: 6)

He estimates, for example, that there is $72.1 billion of dead capital in the urban centres of the Philippines, and $79.4 billion of dead capital in Greater Cairo, waiting to be unleashed through the clarification and formalisation of property relations. While there is little doubt that ambiguous or highly inequitable property rights and tenure regimes create insecurity and constrain the ability of the urban poor to take full advantage of the assets they acquire in informal markets, reforming such regimes is more complicated in practice that de Soto suggests (see Chapter 4).

Informal markets operate where formal markets or the state are absent. For example, they play a large role in the provision of services such as water, transport and electricity. The absence of basic piped water in slums inevitably leads to the emergence of markets for water, provided in plastic sachets and containers or by privately owned tanker trucks, often at inflated prices (see Chapter 5). Where public transport is not provided, informal and privately owned taxi and bus services often fill the gap. And where electricity grids are limited, slum dwellers often purchase power from local dealers who illegally tap into municipal grids and distribute it via ad-hoc networks.

Estimates suggest that the informal economy constitutes more than half of all economic output in countries in low- and middle-income countries (World Bank 2005). While informality is not intrinsically bad, it certainly represents a second-best or least-worst solution from an economic standpoint. Operating in the informal economy may be perfectly rational from an individual's perspective, or it may be a result of deliberate exclusion. Either way, life in the informal economy carries significant risks for individuals and significant costs for society. Ultimately,

> governments have an interest in expanding the net of the formal economy to broaden the tax base, extend the reach of regulations intended to meet important social objectives, and remove distortions in competition between firms in the formal and informal economies. They also have an interest in reducing obstacles to growth faced by firms, and in expanding income earning opportunities for those on the lowest rung of the economic ladder.
>
> (World Bank 2005: 62)

Bringing individuals and small-scale enterprises into the fold of functioning and rewarding urban economies is clearly an important social and economic objective. It is also a complex one, requiring institutional reforms and economic policies that provide positive incentives for individuals and firms to operate along formal lines, as well as strategies that encourage the growth of industries and enterprises that provide employment under decent work conditions. Too often, governments cast a disparaging eye on those operating in the informal economy and attempt to stamp it out, for example, by passing laws and regulations that make street-trading illegal. The extraordinary size of the informal economy in many low- and middle-income countries is probably a transitional phenomenon – a natural by-product of the urban transition – but informality has always been with us. Indeed, if the concept had existed during the Industrial Revolution in Europe or North America, economists would no doubt have observed similarly high levels of informal economic activity. Rather than work against it, governments should instead try to find ways to harness the informal economy and gradually encourage formalisation through positive incentives – not heavy-handed regulation.

Conclusion

There is no question that economic growth spurs urbanisation and urban growth, and there is no question that cities generate wealth. As a nation's economy shifts from subsistence agricultural production to higher value-added sectors, cities become the dominant spaces of economic production and consumption. Through a combination of density and diversity, cities improve the efficiency of economic transactions, improve the productivity of firms and cultivate innovation. But cities also generate costs in the form of congestion and negative externalities, including pollution, health risks and crime. Getting the best from cities requires sound planning and effective economic governance at both local and national levels. Where governments fail to cultivate dynamic urban economies and land, labour and housing markets fail to meet the demands of rapidly growing urban population, informal markets emerge to bridge the gap.

While informal economic activity is not inherently a negative thing, harnessing the full economic potential of cities demands efforts to enhance the security and productivity of those currently operating in the informal economy. As Bairoch (1988) observed:

> [The] contribution of urbanization in low- and middle-income countries today has not wholly mimicked the role of cities in the development of the North. Urbanization has not managed to contribute to improved agricultural productivity in the South, and the gap between information and technology between the North and South encourages the importation of goods, not the production of them. Moreover, the sheer size of cities in low- and middle-income countries today presents unprecedented challenges for policy makers seeking to integrate urban residents into the formal economy of the city.
>
> (Bairoch 1988: 511)

The critical question is whether or not it is possible to tackle historically unprecedented rates of urban poverty while stimulating sustainable economic development in cities in low- and middle-income countries. In the three chapters that follow we will see how neglected social disadvantage, misguided policies, and poor urban planning and governance serve to re-enforce poverty and spatial inequalities in the cities of low- and middle-income countries.

Summary

- **Development is about not only economic growth but also societal change and the diversification of economies, which have important implications for cities. Urbanisation is often an indicator of industrialisation and economic progress.**
- **However, much of Sub-Saharan Africa is rapidly urbanising despite very poor economic growth – a phenomenon often described as 'over-urbanisation'.**
- **Cities are potential engines of economic growth due to the benefits of agglomeration and the various kinds of economies of scale they foster.**
- **Factors such as proximity, high concentrations of skilled labour, the exchange of ideas and diversity of firms in a city contribute to *localisation* and *urbanisation* economies, and can generate *clusters* like the IT sector in Bangalore.**
- **However, there are also considerable costs of associated with urban agglomeration such as congestion, pollution, health risks and crime.**
- **Cities are intimately linked to their rural hinterlands in numerous complex ways; in the 1950s and 1960s various strategies attempted to use cities as 'growth poles' for wider regional development, with limited success.**
- **Globalisation has led to the discourses of 'world cities' and 'global cities', which conceptualise particular cities as critical nodes in the global economy. Linking these ideas back to regional development is the concept of 'global city-regions'.**
- **Cities in low- and middle-income countries are characterised by the large proportion of people engaged in 'informal' economic activity. Debates persist about the extent to which this is a result of deliberate 'exit' or forced 'exclusion' from the formal sector.**
- **Either way, workers in the informal economy face significant costs and risks, and are often deprived of the rights and benefits associated with law and regulation, as well as the security provided by formal institutions such as property rights.**

Discussion questions

1. If urbanisation is usually an indicator of economic development, what accounts for the high rates of urbanisation in Sub-Saharan Africa? Discuss with reference to the concept of 'over-urbanisation'.
2. Why do cities hold such great potential as 'engines' of economic growth?

3 Who gains and who loses in society from:

   (a) external economies of scale
   (b) negative externalities?

4 Why is it so important to understand the various ways in which cities interlink with their rural hinterlands?
5 What are the implications of globalisation for the economic life and role of cities? Discuss both with reference to 'global' cities and 'ordinary' cities in low- and middle-income countries.
6 What are the challenges and opportunities for development posed by the informal economy?

Further reading

Davis, J.C. and J.V. Henderson (2003) 'Evidence on the political economy of the urbanization process', *Journal of Urban Economics*, 53: 98–125.
de Soto, Hernando (2001) *The Mystery of Capital*, London: Black Swan.
Friedmann, John (1986) 'The world city hypothesis', *Development and Change*, 17: 69–84.
Jacobs, Jane (1984) *Cities and the Wealth of Nations*, New York: Vintage.
Marshall, Alfred (1920) *Principles of Economics: An Introductory Volume*, London: Macmillan.
Perry, G.E., W.F. Maloney, O.S. Arias, P. Fajnzylber, A.D. Mason and J. Saavedra-Chanduvi (2007) *Informality: Exit and Exclusion*, Washington, DC: World Bank.
Robinson, Jennifer (2006) *Ordinary Cities*, London: Routledge.
Sassen, Saskia (2006) *Cities in a World Economy*, 3rd edn, Thousand Oaks, CA: Pine Forge Press.
Tacoli, Cecilia (1998) 'Rural-urban interactions: a guide to the literature', *Environment and Urbanization*, 10(1): 147–166.

Useful websites

Global Development Research Centre (GDRC) urban informal sector page: www.gdrc.org/informal/index.html
International Institute for Environment and Development (IIED) rural–urban linkages page, with links to numerous publications: www.iied.org/HS/themes/Rural–Urbanlinks.html
International Labour Organization (ILO): www.ilo.org

4 Urban poverty and vulnerability

Introduction

Persistent poverty is the antithesis of progress – and poverty is disturbingly persistent. After more than 50 years of concerted efforts to reduce global poverty, limited progress has been made. And while it is widely held that economic development is a necessary condition for poverty reduction, it is insufficient on its own: pockets of poverty can be found in every country of the world, rich and poor. Why, and what can be done about it? These questions have motivated decades of research into the nature, causes and extent of poverty in rich and poor countries alike. They have also inspired the growth of the international aid industry, which, since the 1950s, has funnelled trillions of dollars into efforts to promote growth and alleviate poverty. While some gains have been made in the fight against global poverty, it remains a disturbingly widespread phenomenon. Furthermore, although poverty in low- and middle-income countries has generally been portrayed as a rural phenomenon, recent evidence suggests that economic growth and rapid urbanisation are contributing to a decline in overall poverty but a rise in urban poverty.

In this chapter we begin by examining the ways in which poverty has been conceptualised and measured in the past. Definitions and metrics are important for identifying the location and scope of poverty, but can obscure important differences among those classified as poor. It is important to recognise, for instance, that the manifestations of poverty are somewhat different between rural and urban contexts. Furthermore, intra-urban inequalities are often overlooked in aggregate statistics, thereby underestimating the scale of urban poverty, which is a multidimensional phenomenon. Insecure livelihoods, poor housing conditions, lack of access to basic services, social fragmentation and exposure to crime and violence conspire to create conditions of relentless vulnerability for the urban poor. In this chapter we focus specifically on the two critical aspects of poverty in cities: livelihoods and shelter. Efforts to combat urban poverty with employment and income-generation schemes have met with limited success, due in part to the complex nature of urban livelihoods. This becomes particularly clear when we consider housing, which is both an essential component to a sustainable livelihood strategy and an economic asset. Inadequate shelter exposes millions of poor to environmental hazards, while insecurity of tenure renders millions of urban residents vulnerable to eviction and undermines the value of their shelter as an asset. The sheer size of the global slum population today attests to the fact that policy responses to housing shortages have also met with very limited success. We continue our exploration of urban poverty and vulnerability in Chapters 5 and 6, which explore issues related to managing urban environments and ensuring human security in a rapidly urbanising and globalising world.

Understanding poverty: definitions, measurement and trends

Poverty can be defined most simply as a lack of one or more of those things that determine quality of life. But what are 'those things' and how do we define quality of life? Over time, conceptualisations and measurements of poverty have evolved from a concern with subsistence, to a focus on basic needs, to the notion of relative poverty or deprivation. The notion of absolute or subsistence poverty, which dates back to the nineteenth century, defines poverty in relation to the minimal physical requirements of human beings, such as food, clothing and shelter. To be poor, from this perspective, is to lack the basic means to survive. The most common way of measuring poverty thus understood is to measure the number of people who fall below a specified level of income, usually called the poverty line, to get what is called a poverty headcount measure. There are a variety of ways of calculating poverty lines. Some rely on income thresholds (e.g. an income of US$1 a day or less would qualify an individual as poor), others use a basket of consumption (e.g. the inability to purchase a pre-specified set of essential goods would classify an individual as poor), and still others on caloric intake (e.g. a daily intake of less than some predetermined number of calories would classify an individual as poor). The analytical strength of this approach to poverty assessment rests in the clarity of definition, allowing ease of measurement through identifiable indicators such as income. In policy terms it provides a clear basis for setting the threshold of, for example, a minimum wage or targeted income transfers such as welfare grants.

Nevertheless, this approach has its shortcomings. For example, if 40 per cent of a country's population falls below the poverty line, we have no idea of where particular individuals fall – are their incomes close to the poverty line or close to zero? Furthermore, if resources are targeted so that they reach the poorest of the poor rather than those closer to the poverty threshold, the result may be an improvement in the lives of the least well-off, but the aggregate number of poor people living below the poverty line (i.e. the poverty headcount) may not shift at all. These kinds of indices can also mask differences in the way poverty is experienced. If we take the example of measuring poverty across rural and urban areas, calories are purchased in cities rather than grown so that a larger proportion of monetary income goes towards basic necessities such as food, shelter, even water, than in rural areas. As such, poverty headcount measures often underestimate the extent of urban poverty – even when they do attempt to correct for the differing ways in which people acquire necessities. There are also the usual problems associated with gathering accurate data, especially given that people tend to be reticent about revealing their true economic status.

It was in part as a result of the shortcomings of poverty lines and the absolute poverty approach that a focus on basic human needs gained prominence in the 1970s. It was argued that policies directed at addressing minimum requirements for basic consumption were inadequate and that in addition to food, shelter and clothing people needed support in making a living as well as access to decent services, healthcare and education. The critique also recognised that human beings are more than machines and have more complicated needs than the means of sustenance and reproduction, including for example, social and psychological needs. While compelling, the basic needs approach suffered conceptually from vague definitions and the application of inconsistent criteria for establishing basic human needs. It was difficult to establish a person's essential needs and how value might be placed, for example, on dignity, respect and relationships of love or friendship. In policy terms the basic needs approach offered income-generating projects and sought to deliver basic services such as safe water and adequate sanitation, primary education and basic healthcare. While admirable, it did not distinguish between one context and another and did little to identify or address the underlying causes of poverty and inequality.

One reason for this neglect is that absolute poverty lines, based on some minimum level of income or basket of goods, reveal nothing of income distribution within a context or group. A corrective was the notion of *relative poverty*, with a person understood to be in relative poverty when his or her income fell below that of others in a particular country. Relative poverty measures were developed with the poverty line being based, for example, on half the medium income or a fraction of average national income. The advantage of a relative approach is that it captures inequality and hence gives us a clearer sense of relative deprivation within a country. Relative deprivation can also be assessed by identifying whether people lack the minimum requirements for both material and social needs in a particular time or place. According to Townsend (1993):

> People are relatively deprived if they cannot obtain, at all or sufficiently, the conditions of life – that is, the diets, amenities, standards and services – which allow them to play the roles, participate in the relationships and follow the customary behaviour which is expected of them by virtue of their membership of society. If they lack or are denied resources to obtain access to these conditions of life and so fulfil membership of society they may be said to be in poverty.
> (Townsend 1993: 36)

For example, in many Asian societies, social interaction is predicated on being able to offer visitors to the home a cup of tea. Tea has little nutritional or caloric value but not to be able to offer it excludes people from social participation according to customary behaviour. The relative deprivation approach to poverty is echoed in Amartya Sen's work. Sen (1999: 87) defines poverty as *capability deprivation*, by which he means the condition of lacking one or more of the capabilities that allow people to live the kind of life they have reason to value.

The relative deprivation and capability deprivation perspectives offer an important corrective to simplistic conceptualisations of poverty based only on income or consumption measures because they concede that social and political factors are often important in determining the experience of poverty. For example, two people – a man and a woman – may have precisely the same income, but the woman may have less choice about how she earns or spends her income, or may not be allowed to express personal or political opinions if she lives in a society that persistently discriminates against women. In this case, although the man and the woman may appear to have equal access to financial resources, the socio-political context renders the woman poorer than the man from a relative deprivation perspective.

Efforts to measure poverty have evolved roughly in line with these changing conceptualisations of poverty. Generally, however, reliance is still placed on subsistence poverty measures that use indicators such as income, caloric intake or other proxies for consumption, particularly for cross-national comparison. Perhaps the most well-known poverty index is the World Bank's 'dollar-a-day' measure, which defines extreme poverty using a standard income threshold. Those who live on less than the equivalent of one-dollar-a-day are said to live in extreme poverty; those living on less than two-dollars-a-day are classified as 'moderately poor'. Many international agencies still use this measure as an international index of poverty. Criticism of such arbitrary thresholds for classifying poverty contributed to the creation of the Human Development Index (HDI), which was developed by UNDP in the early 1990s under the guidance of an international group of scholars drawing on the idea of capability deprivation. The HDI is a composite indicator of development that factors in life expectancy, educational attainment and per capita income. Although crude, it is a more holistic indicator of development than per capita income alone and comes closer to capturing relative capability deprivations – in this case literacy and life expectancy are used as rough indicators of basic human capabilities (see Box 4.1).

Box 4.1

Measuring human development

There are a variety of ways of defining and measuring poverty. The most common strategy is to establish a 'poverty line' or 'poverty threshold' based on income. For example, in the United States a family of four is classified as 'poor' if annual household income falls below US $21,000. For international comparisons, the World Bank uses a standard one-dollar-a-day measure to calculate the percentage of a nation's population that is 'very poor.' The Human Development Index, developed by the United Nations Development Programme, takes a different approach, informed by Amartya Sen's (1999) 'capability deprivation' perspective. The HDI is a composite indicator that factors in levels of education (as reflected by literacy rates and school enrolment rates), and life expectancy as rough proxies for basic human capabilities. The UNDP's annual *Human Development Report (HDR)*, which publishes the HDI, goes further by providing a wide range of social development indicators related to gender equality, employment, human rights, and security (to name a few). While human development (as measured by the HDI) and income levels are highly correlated, the HDI does offer a slightly different interpretation of levels of development. The table below provides some examples of where countries rank on the HDI versus how they rank based on a standard per capita income scale of development.

Country	*HDI rank*	*GDP per capita rank*	*GDP rank – HDI rank*
Iceland	1	5	4
Norway	2	3	1
Australia	3	16	13
United States	12	2	–10
United Kingdom	16	11	–5
Chile	40	55	15
Cuba	51	94	43
Brazil	70	67	–3
China	81	86	5
Philippines	90	101	11
Algeria	104	82	–22
South Africa	121	56	–65
India	128	117	–11
Angola	162	129	–33

Source: UNDP (2007)

From this selection of countries we can see that HDI rankings are often very close to income per capita rankings of development. However, there are some notable differences. For example, the United States has the second highest level of income per capita (after Luxembourg), but ranks twelfth on the HDI index. Conversely, Cuba ranks ninety-fourth by income, but due to high-quality education and healthcare services ranks much higher on the HDI. By contrast, South Africa, which is burdened by the scourge of HIV/AIDS, has a very low life expectancy despite middle-income status. Finally, the oil-exporting countries of Angola and Algeria have much higher levels of income per capita than human development.

The reason indices such as these matter is because there is strong congruence between how poverty is conceptualised and measured and how it is addressed. This can be illustrated by comparing the World Bank's *World Development Report (WDR) 1990* and its *WDR 2000*, both of which had poverty as their theme (World Bank 1990, 2000). The *WDR 1990* was driven by a conceptualisation of poverty based on a subsistence approach, where poverty was defined as 'the inability to attain a minimum standard of living' (World Bank 1990: 26). The influence of the basic needs approach was also evident in that the report focused in addition on shortfalls in education and health. The policy recommendations involved a three-pronged approach to poverty reduction: economic growth, social safety nets for the very poorest and human development through improvements in health and education.

If we turn to the *WDR 2000* we can observe a broadening of the World Bank's perspective illustrated by the following extract from the report:

> Poverty is pronounced deprivation in well-being. But what precisely is deprivation? The voices of poor people bear eloquent testimony to its meaning. To be poor is to be hungry, to lack shelter and clothing, to be sick and not cared for, to be illiterate and not schooled. But for poor people, living in poverty is much more than this. Poor people are particularly vulnerable to adverse events outside their control. They are often treated badly by the institutions of state and society and excluded from voice and power in those institutions.
>
> (World Bank 2000: 15)

There are continuities of analysis between the two reports, and growth remains centre stage as *the* main solution to problems of poverty, along with the provision of basic services. However, it is acknowledged here that simply providing resources to poor people, whether money or services, is not sufficient and other factors enter the equation. Institutional reform needs to accompany economic growth in recognition of the fact that there are institutional barriers that prevent the poor from accessing public goods, services and decision-making arenas. It is noted too that vulnerability matters, whether due to factors such as drought, national price hikes or the effects of international capital flows. Lastly, inequities are recognised, not only across society as a whole but within deprived groups as well, such as those based on gender, ethnic, racial or religious differences.

What accounted for this change? The relative deprivation approach not only inspired a more comprehensive set of indicators for measuring poverty, it also changed the way it was analysed. This included drawing on the expertise of a wider range of specialists both from within and outside the discipline of economics. For example, anthropologists and sociologists offered important insights into poverty processes, highlighting the way in which people fall into and escape from poverty over their lifetimes, and how intergenerational dynamics affect socio-economic mobility. One example of how such approaches were mainstreamed was the Voices of the Poor (VoP) project (Narayan *et al.* 2000a, 2000b). This was a World Bank initiative that informed the analysis adopted in the *WDR 2000* by providing an in-depth exploration of the multidimensional nature of poverty through qualitative analysis. The VoP project sought to understand the nature of poverty by talking to those who actually experience it; by collecting, cataloguing and analysing interviews with 60,000 people across 60 countries. Holistic approaches to poverty research such as this helped to reshape approaches to poverty reduction and the international development agenda more generally.

This is apparent in the content of the Millennium Development Goals (MDGs), which were adopted by 189 countries following the September 2000 United Nations Millennium Summit, where world leaders agreed to a set of time-bound targets for combating poverty, hunger, disease, illiteracy, environmental degradation and discrimination against women (see Box 4.2). This comprehensive list of objectives has helped to focus the attention of governments, donors and development professionals on

Box 4.2

The Millennium Development Goals

The Goals	*Implementation*
1 Eradicate extreme poverty and hunger	✓ Reduce by half the proportion of people living on less than a dollar a day
	✓ Reduce by half the proportion of people who suffer from hunger
2 Achieve universal primary education	✓ Ensure that all boys and girls complete a full course of primary schooling
3 Promote gender equality and empower women	✓ Eliminate gender disparity in primary and secondary education preferably by 2005, and at all levels by 2015
4 Reduce child mortality	✓ Reduce by two thirds the mortality rate among children under five
5 Improve maternal health	✓ Reduce by three quarters the maternal mortality ratio
6 Combat HIV/AIDS, malaria and other diseases	✓ Halt and begin to reverse the spread of HIV/AIDS
	✓ Halt and begin to reverse the incidence of malaria and other major diseases
	✓ Integrate the principles of sustainable

Goal	Targets
	development into country policies and programmes; reverse loss of environmental resources
7 Ensure environmental sustainability	✓ Reduce by half the proportion of people without sustainable access to safe drinking water ✓ Achieve significant improvement in lives of at least 100 million slum dwellers, by 2020
8 Develop a global partnership for development	✓ Develop further an open trading and financial system that is rule-based, predictable and non-discriminatory. Includes a commitment to good governance, development and poverty reduction – nationally and internationally ✓ Address the least developed countries' special needs. This includes tariff- and quota-free access for their exports; enhanced debt relief for heavily indebted poor countries; cancellation of official bilateral debt; and more generous official development assistance for countries committed to poverty reduction ✓ Address the special needs of landlocked and small island developing states ✓ Deal comprehensively with developing countries' debt problems through national and international measures to make debt sustainable in the long term ✓ In cooperation with the developing countries, develop decent and productive work for youth ✓ In cooperation with pharmaceutical companies, provide access to affordable essential drugs in developing countries ✓ In cooperation with the private sector, make available the benefits of new technologies—especially information and communications technologies

Source: UNDP (2003)

the diverse needs of vulnerable people. Unfortunately, this laudable set of goals will very likely remain unfulfilled by the 2015 deadline that was set for achievement, not least because the means of achieving these goals remain unclear. Indeed, one of the key critiques levelled against the MDG framework is the absence of any significant articulation of the *causes* of poverty – a point we will explore in more depth below.

Another noticeable omission from the MDGs is any reference to the *spatial* dynamics of poverty – apart from the reference in Goal Seven to improving the lives of 100 million slum dwellers by 2020. Tackling poverty requires a firm understanding of not only the causes of poverty, but also the location of poverty. On the one hand, public policies and aid interventions cannot be effectively targeted at the poor unless their location is known. On the other hand, where someone is located affects his or her experience of poverty, which has important implications for those seeking to devise effective interventions.

By any measure, the majority of those categorised as poor still live in rural areas. However, it is increasingly apparent that the location of poverty is rapidly shifting from rural to urban areas as low- and middle-income countries become increasingly urbanised. Moreover, in the past urban poverty has probably been underestimated due to data constraints and the nature of measurement (Haddad *et al.* 1999). Nevertheless, increasing data availability and more in-depth analyses of intra-urban inequalities reveal a clear trend. World Bank research indicates that between 1993 and 2002 the number of people living on one-dollar-a-day or less fell by 150 million in rural areas but rose by 50 million in urban areas, suggesting that 'The poor have been urbanising even more rapidly than the population as a whole' (Ravallion *et al.* 2007: 1). There is some dispute about how long it will take before a *majority* of the global poor live in urban areas, but it is clear in which direction things are moving: 'Poverty is clearly becoming more urban' (Ravallion *et al.* 2007: 27).

This recent global analysis of poverty trends based on a simple income-based measure is corroborated by more in-depth studies that have examined other aspects of urban poverty. For example, a study drawing on extensive health data from 14 countries concluded that:

> the number of urban poor is increasing; the share of the urban poor in overall poverty is increasing; the number of underweight preschoolers in urban areas is increasing; and the share of urban preschoolers in overall numbers of underweight preschoolers is increasing. The locus of poverty and undernutrition does seem to be changing from rural to urban areas.
>
> (Haddad *et al.* 1999: 1897)

In the case of Dhaka, Bangladesh, Jane Pryer has observed that 'although urban populations overall have lower rates for infant, child and adult mortality, compared with rural populations, the urban poor suffer very high rates indeed compared with either group' (Pryer 2003: 83). This finding highlights the problem with using aggregate statistics that categorise poverty only in terms of 'rural' and 'urban' without taking into account intra-urban disparities. As Haddad *et al.* (1999) observed:

> Too often, all urban households – rich and poor – are averaged out to provide one single estimate of poverty or malnutrition. In countries with high income and social inequalities this can be particularly misleading.
>
> (Haddad *et al.* 1999: 1899)

Beyond the problems associated with accurately measuring the scale of urban poverty, analyses that focus only on quantifiable indicators fail to capture the complexity of urban poverty as experienced by the urban poor. Although low incomes and ill-health are surely signs of deprivation, effectively addressing urban poverty requires an understanding of the insecurity and vulnerability experienced by poor people in relational as well as material terms. It is persistent vulnerability, perhaps more than absolute deprivation, which characterises urban poverty and prevents the urban poor from enhancing their capabilities and getting a foothold on the socio-economic ladder (Moser 1996). Moreover, vulnerability has its roots as much in social and institutional exclusion as in material want.

Urban poverty and vulnerability

We have to be careful not to draw the distinction between rural and urban poverty too rigidly, given that many drivers of social disadvantage operate at national and even the international level (Wratten 1995). Nevertheless, if we accept that poverty is more than absolute and relative deprivations – which are static concepts – and is also a dynamic process, then we can identify *processes* of impoverishment that are particularly urban in nature. In cities, people can become caught in a cycle of poverty that is a direct result of the myriad vulnerabilities that they face. Box 4.3 summarises the key characteristics of urban poverty and vulnerability. Note the difference between this list and the list of MDGs in Box 4.2. The MDGs emphasise the manifestations of poverty (i.e. low incomes, poor health), whereas the characteristics listed in Box 4.2 emphasise the vulnerabilities experienced by the urban poor.

Box 4.3

Key characteristics of urban poverty and vulnerability

- Reliance on a monetised economy
- Reliance on employment in the informal economy
- Poor quality housing
- Insecurity of tenure (for both owners and tenants)
- Lack of access to basic infrastructure and affordable services
- Susceptibility to diseases and accidents
- Environmental hazards, including the impacts of natural and man-made disasters
- Social fragmentation
- Exposure to violence and crime and fear of violent crime
- Increasing exposure to warfare and terrorist attacks.

In cities, vulnerability is first and foremost linked to people's reliance on a monetised economy. In other words, urban residents need money to acquire such basic necessities as food, water, shelter and clothing. Absolute reliance on a monetised economy is one of the key differences between urban as compared to rural poverty – rural residents can also rely on subsistence farming, or even foraging and hunting in lean times, while urban residents cannot. Although there is a growing body of research on the possibilities for urban agriculture to contribute to livelihood strategies, it is unlikely to significantly meet the income and consumption needs of the majority of poor people, especially in large and mega-cities. As agricultural output falls short of demand for food, it is in cities that hunger will become a growing problem. This is already evident from the growing number of food riots in cities of low- and middle-income countries (see Box 4.4).

Operating in a primarily monetised economy makes wage labour the centrepiece of urban livelihood strategies, alongside self-employment and small-scale enterprise activities. Indeed, a study of poverty and governance across nine cities of Africa, Asia and Latin America found 'jobs and income earning opportunities . . . to be the most fundamental preoccupations of the urban poor' (Beall 2004: 54). This is hardly surprising, given the dependence on cash for survival in urban areas. However, the paucity of job opportunities in the formal economy means that most people in low- and middle-income cities depend on employment in the informal economy (as noted in Chapter 3). Informal employment is

Box 4.4

Urban food riots in 2007–2008

During 2007 and 2008 the world witnessed spontaneous protests and riots in over 30 cities in countries across the world, including Guyana, Haiti and Bolivia in Latin America and Caribbean; Cameroon, Egypt, Côte d'Ivoire and Senegal in Africa; Surinam, Bangladesh, South Korea and Uzbekistan in Asia; and Yemen in the Middle East. In Cameroon food riots resulted in 24 deaths, while in Haiti six died and the prime minister was forced to step down. In Côte d'Ivoire anti-riot police used tear gas to disperse the predominantly female demonstrators and the president cancelled customs duties and cut taxes on basic household products after days of violent protests in the capital, Abidjan.

This wave of protest and violence followed a steep rise in prices for staple foods over just a few months – the prices for wheat and corn rose by 68 per cent in the first quarter of 2008. This price shock was the result of a confluence of factors, including high oil prices (which affect the cost of fertilisers and transport), rising demand for bio-fuels (which encourages farmers to divert production away from food and toward bio-fuel feedstock), and the growth of demand for food and other primary products from the expanding economies of Asia and Latin America.

The Executive Director of the World Food Programme warned that the world was confronting 'a new face of hunger' and that 'often we are seeing food on the shelves but people being unable to afford it'. Increasingly this is an urban face as people have started to make their misery known. Governments and international development agencies have geared up to provide better support to agricultural development and food production. It remains to be seen whether they will pay similar attention to the plight of the urban poor.

This unrest in cities over rising food prices is the angry and desperate face of urban poverty, one that is evident in the history of advanced economies in the past, and may well persist globally into the future.

Sources: BBC News (2008); Patel (2008); Pilkington (2008)

insecure by definition. Without enforceable contracts, workers in the informal sector have no assurances of job security.

For the most disadvantaged – often women and children – insecurity of employment may be less of a concern than the hazards associated with available employment. Many urban dwellers rely on livelihoods that put them at immediate physical risk – such as hawking in busy traffic areas, retrieving and recycling dangerous waste products, or engaging in sex work. In Addis Ababa, for example, an incredible 78 per cent of sex workers are HIV-positive in high-risk areas. There is ample evidence to demonstrate not only higher concentrations of HIV/AIDS in urban areas but also higher prevalence and infection rates (van Donk 2006: 157) – a consequence of the density of populations in urban areas. And it is usually the poor who lack access to contraception and to the anti-retroviral drugs that HIV-positive patients require. Furthermore, the immune systems and general health of the poor are likely to be compromised due to poor nutrition, making it increasingly likely that HIV will become full-blown AIDS. To make things worse, cramped and overcrowded housing conditions significantly increase the probability of falling prey to a secondary illness that can result in death. For instance, those infected with HIV are already 100 times more likely to contract tuberculosis than the non-infected. In Nairobi today there are more tuberculosis cases among its population of 3 million people than in the entire United States, while Kibera, the city's largest slum, is thought to house more than 50,000 AIDS orphans (National Public Radio (NPR) 2002).

The insecurities associated with operating in a monetised economy and dependence on informal employment are exacerbated by insecure housing arrangements. As noted in Chapter 1, a large proportion of urban dwellers in low- and middle-income countries live in slums and informal settlements, which emerge where formal housing markets and government housing programmes fail to keep pace with urban growth. For example, in India more than 50 per cent of all urban residents live in slums; in Peru more than 70 per cent; and in Uganda more than 90 per cent of the urban population live in slums (UN-Habitat 2006). Housing in these areas generally consists of makeshift shelters of corrugated iron or zinc sheets, scavenged pieces of cardboard and wood, and industrial scraps. Even those living in apparently more permanent structures often fall prey to the poor quality of their dwellings. For example, the poor quality of dwellings in Bam in Iran has been identified as the primary cause of the 40,000 lives lost there in an earthquake in 2003 (UN-Habitat 2006: 136). Overcrowding, poor construction, bad locations and dwelling density also render informal settlements prone to accidental damage such as flooding and fire. Fire in particular is common in informal settlements due to people cooking and lighting with naked flames.

Many, if not most, of the urban poor live in a state of legal ambiguity without clearly defined or enforced tenancy rights. Squatters (i.e. those who illegally occupy public or private property in and around cities) are tolerated only so far as there is low demand for the land they have occupied. Tenants often face rent hikes and harassment on the part of formal and informal landlords. Without legal titles, or legally binding rental or lease agreements, squatters and tenants remain vulnerable to expropriation by land speculators, developers, and government agents. When the interests of land developers or public agencies collide with those of informal settlement residents, they can be reclassified (from 'informal' to 'illegal'), forcibly evicted, and their homes demolished – a process that often incites violent confrontations. According to the Centre on Housing Rights and Evictions (COHRE 2006), globally over 4.2 million people were forcibly evicted from their homes between 2003 and 2006 – most of them in urban areas. Across Africa, nearly 2 million people were evicted from their homes between 2003 and 2006; in Asia the figure is just over 2 million for the same period; and millions more continue to live with the threat of eviction (COHRE 2006). The widespread prevalence of the problem led UN-Habitat (the United Nations agency dedicated to urban development issues) to make Security of Tenure one of its two key programmatic themes.

The indifferent and sometimes hostile attitudes that prevail towards informal settlements and slum dwellers contribute to the under-provision of basic services. Dwellings in such communities are rarely incorporated into formal urban infrastructure networks, such as water, drainage and sewerage systems. Limited access to these basic services exposes poor urban residents to debilitating diseases. Almost half of all urban residents in Africa, Asia, and Latin America are victims of ill-health associated with poor water and sanitation facilities, such as diarrhoeal diseases and worm infections. Approximately 5 million people in rural and urban areas die every year from diseases carried in contaminated water (Sacquet 2002) and 3 million of these deaths can be attributed to one of the world's most easily treated ailments: diarrhoea. The failure to address the problem is due as much to lack of political will and institutional barriers as to inadequate resources and poor technology (see Figure 4.1). Such risks to personal health are worsened by the location of informal settlements. Poor urban residents often settle on marginal land – land that is of little value to developers because of its undesirable topography or location. Such settlements might be located on flood plains, riverbanks, steep slopes and reclaimed lands, which are particularly susceptible to natural disasters. In 1998, floods and landslides caused by Hurricane Mitch claimed the homes and lives of hundreds of urban dwellers in Latin America (UN-Habitat 2006: 136). As we show in Chapter 5, it is probable that climate change will make things worse.

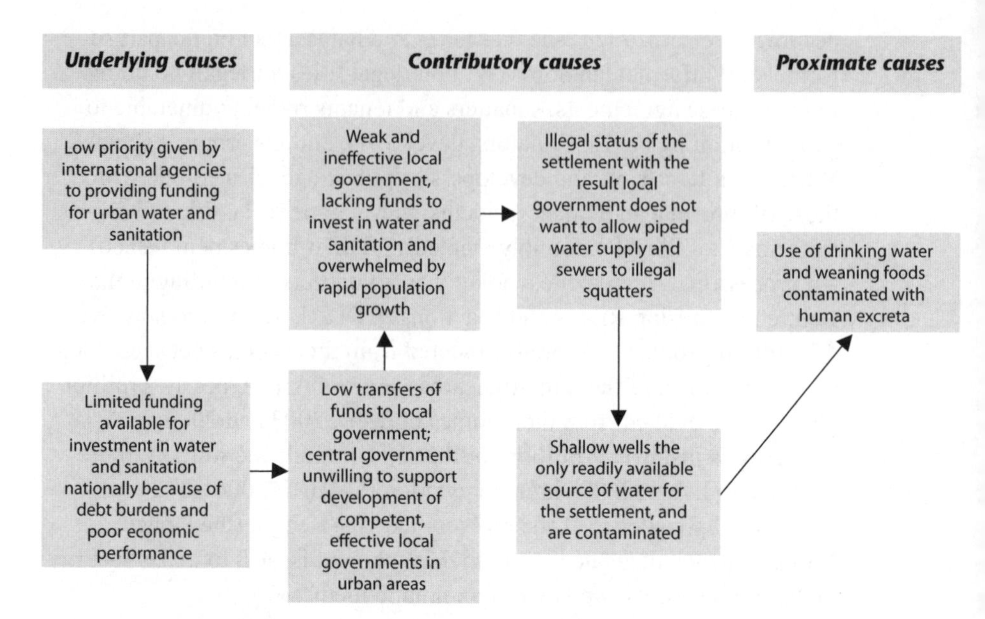

Figure 4.1 Underlying, proximate and contributory causes of diarrhoeal diseases in squatter settlements

Source: United Nations (2003)

Apart from the obvious consequences of economic and environmental vulnerability, the urban poor face more subtle forms of insecurity such as the fragmentation of social networks of trust and reciprocity. Kinship ties, social networks and connections with friends and neighbours often serve as a kind of informal social safety net for the poor, but they are also demanding of people. It is hard to reciprocate when you have nothing to share. Customary forms of managing health risks, economic insecurity and other forms of vulnerability are placed under severe stress when low-income urban dwellers face hard times, and the struggle to survive in cities can lead to the breakdown of families and friendships. Borrowing and begging by destitute kin, with no possibility of repayment, drains limited resources and becomes problematic for relatives who themselves are already on the edge (Beall 1995, 2001). The strains of maintaining reciprocal relationships under conditions of persistent vulnerability can contribute to social fragmentation and alienation. In some cases, families may find themselves forced to sell what assets they have, to scavenge in the street or even to engage in illegal activities rather than borrowing from relatives and friends to survive hard times.

Lastly, in identifying the characteristics of urban poverty, it is important to highlight that the experience of vulnerability and deprivation is frequently shaped by conflict and violence both within and beyond cities. As we show in Chapter 6, social fragmentation and deprivation can encourage socio-economic forms of violence (such as crime and gang violence) within cities, but even larger conflicts taking place beyond city boundaries can have important effects on life in cities. Under conditions of war, people often flock to urban centres when life and livelihoods in villages become unviable, rapidly altering the composition of urban populations. Refugees and local communities are thrown together and forced to compete for urban resources, services and limited income-generating opportunities (Beall and Esser 2005; Beall and Schütte 2006). Cities and urban populations are also increasingly targets of politically motivated violence in the form of terrorism (Beall and Graham 2004a).

As cities in low- and middle-income countries continue to expand, the number of people caught in webs of vulnerability created by these myriad forces continues to rise. In what follows we focus on two critical policy arenas through which poverty and risk in cities and towns can be mitigated: the securing of sustainable urban livelihoods and access to adequate and secure housing.

The challenge of sustainable livelihoods

In the 1950s and 1960s, urban poverty was largely perceived as a temporary phenomenon – a product of the disruptive process of industrial transition. It was generally believed that the urban poor would eventually be absorbed into formal labour markets. Urban development policies focused primarily on investments in infrastructure and housing for the formal workforce and their families. By the 1970s, however, it became clear that urban labour markets were failing to keep pace with urban growth, evident from expanding slums and urban informal economies. The discourse of international development began to shift towards 'basic needs', with increasing attention being paid to such issues as access to primary healthcare, education and other essential services, as well as the need for individuals to have access to employment and income earning opportunities and the right to participate in community life (Stewart 1985; Streeten *et al.* 1981). But this focus on 'redistribution with growth' was disrupted by economic crisis and the shift toward market-oriented solutions to development in the 1980s. Under the rubric of structural adjustment programmes, proactive government efforts to generate employment and provide social services generally gave way to policies designed to help poor people help themselves, for example through small and medium enterprise development and microfinance initiatives.

The 'rolling back' of the state under structural adjustment reduced social sector spending in many countries, exacerbating the already vulnerable position of the urban poor:

> The urban poor were faced with a price-income squeeze, as the effects of unemployment and downward pressure on wages were compounded by the marketization of public goods. The majority of new recruits to the labour market were left with underemployment in the informal sector as the only option left open to them.
>
> (Watt 2000: 103)

Moreover, the impact of job losses, the withdrawal of food and transport subsidies and the introduction of user charges for goods and services fell more heavily on some than others, with women being particularly vulnerable to downward pressures on income and consumption and children engaged in income-generating activities in greater numbers (Moser 1996) (see Plate 3). In response to the negative social impacts of macroeconomic reform and a barrage of critiques from scholars and

Plate 3 Wastepickers in Bangalore, India

Source: Jo Beall

development practitioners, new policies and programmes were devised to ease the pain of adjustment. The focus shifted from an orientation towards direct poverty alleviation with the basic needs approach in the 1970s, to a focus on 'economic growth and trickle down in the 1980s', to a 'renewed interest and attention to social issues and poverty reduction' (Kanji 2002: 234).

Calls for 'adjustment with a human face' (Cornia *et al.* 1987) were met with social development initiatives that were intended to serve as temporary social safety nets, primarily for those made unemployed through cut-backs in public sector employment. The main objectives included improving social and economic infrastructure, job creation (generally temporary), community development to enable communities to manage resources themselves, improved delivery of social services and strengthening of municipal government capacities (Kanji 2002: 236). A common vehicle was through social funds, an early example of which was the Programme of Action to Mitigate the Social Cost of Adjustment (PAMSCAD) devised and implemented in Ghana in the late 1980s. The objective of PAMSCAD was 'to generate works (and incomes) through the construction of basic needs facilities and provide both work and training for retrenched workers in rural and urban deprived areas' (Donkor 2002: 226). Although the programme did manage to generate employment and public works for a short period, it quickly withered away. In the end 'it neither mitigated poverty nor did it indeed have the capacity to do so' (Donkor 2002: 227). Initially intended as temporary, social funds have persisted as a first choice strategy for channelling resources towards social development, but they have come in for considerable criticism on the grounds that they are not good at targeting those most in need, can be used as a political tool both to punish and reward, are dependent on outside funding and are not integrated into government organisations, and are therefore ultimately unsustainable (Jørgensen and Van Domelen 2001; Tendler 2000).

The impact of social funds specifically on urban poverty reduction has been limited as well, although there have been some success stories, particularly when their promotion has been linked to local government processes and existing agencies for urban development. For example, funding from the World Bank and the ILO led to the establishment of Agences d'Exécution de Travaux d'Intéret Public (AGETIPs) in a number of Francophone African countries, agencies with a strong emphasis on poverty reduction through labour intensive infrastructure provision by way of public–private partnerships (PPPs). Despite such achievements at project or programme level, however, social funds are inadequate to the task of addressing urban poverty writ large (Beall 2005).

The failure of successive policies to make a significant dent in urban poverty over the past several decades led to a shift in thinking about how to deal with disadvantage at the urban scale. A key critique of past approaches is that they focused either on income generation or ensuring minimum consumption levels, neither of which proved a sustainable way of lifting people out of poverty (Moser 2007: 87). Instead researchers have increasingly sought to expose the complexity of urban livelihood strategies and policy makers have begun to address the challenges of establishing sustainable livelihoods that reduce urban vulnerability (Rakodi 2002). Based on years of research on agrarian change and influenced by Amartya Sen's analysis of poverty and deprivation (Chambers and Conway 1992), the concept of 'livelihoods' has come to the fore. Livelihoods are not just about income earning but rather the wide spectrum of capabilities, assets and activities required to make a living. Commonly known as the sustainable livelihoods framework (SLF), this perspective adopts a broad conceptualisation of poverty and vulnerability and is associated with integrated approaches to development that include a focus on economic opportunity, social protection and a concern with environmental sustainability (Scoones 1998). The key insight of the SLF is that income from waged employment is only one dimension of the livelihood strategies of most individuals and households. Other factors, such as achieving material assets, good health, skills and knowledge, and maintaining social networks constitute equally important components of people's livelihood strategies. A livelihood strategy is deemed sustainable when it allows an individual or family to cope with and recover from stresses and shocks in the immediate term and in the long-term, as well as maintain or enhance the capabilities and assets of a household and its members, both now and in the future, without undermining the natural resource base (Carney 1998: 4). There are a number of options available to low-income households in managing their livelihoods. These include reducing consumption, increasing labour or income-earning activities and investing in tangible and intangible assets, such as houses and social networks respectively. In order to ensure long-term security households might substitute one asset for another, for example selling jewellery to purchase tools; or sacrificing assets for consumption to deal with short-term survival needs under conditions of stress.

The SLF has been taken up in different forms by national governments, international NGOs such as Oxfam and Care International, bilateral donors such as DFID and by UNDP. As a framework for analysing poverty, the SLF is useful because it invites a holistic approach, but as an operational tool it has proved more limited. SLF-informed initiatives frequently encompass fairly traditional projects and programmes, such as employment-generating infrastructure projects, microfinance programmes and the promotion of participatory approaches to development planning and practice. However, SLF-informed development strategies often combine such initiatives in creative ways that are more attentive to the challenges of generating sustainable improvements in livelihoods.

The SLF was first developed in relation to rural livelihoods but was subsequently explored in relation to urban risk and vulnerability with a particular focus on asset accumulation (Moser 1996, 1998; Rakodi 1999). Pioneering has been Caroline Moser's work, which focuses on poverty dynamics and mobility, including through intergenerational resource accumulation and loss. Drawing on the work of Anthony Bebbington (1999) and Amartya Sen (1981), she defines assets as a stock of financial, human, natural or social resources or capital endowments on which people can build livelihoods and which also gives them the capability to be and to act (Moser 2007: 2) (see Box 4.5). She shows how urban households manage complex asset portfolios over their life course, arguing that intergenerational strategies are not only about household survival on a day-to-day basis but about investment in long-term security (Moser 1996, 1998, 2007).

While frameworks such as the assets and sustainable livelihoods framework can provide useful principles by which to organise our understanding of social reality, they can also become rigid grids that seek to codify complexity and do not easily accommodate the messiness and micropolitics of people's everyday home and working lives. A second weakness is that the livelihoods perspective tends to emphasise agency – what the poor can do for themselves – over the inalienable structural factors that condition and limit the extent of people's agency (Beall 2002: 75). Clearly both perspectives are important. To illustrate this, take the example of the urban informal economy (discussed in Chapter 3). Hernando de Soto (1989) endorsed informality as 'the other path' by which poor people, as determined agents of their own destiny, find ways to circumvent the rules, regulations and obstacles put in place by inefficient and uncaring governments. Policy makers were swift to place faith in and rely upon the creative energy and rugged self-reliance of the working poor. However, informalisation is as much a structural issue as one of agency, serving multiple and asymmetrical interests, tied in turn

Box 4.5

Livelihood assets or 'capitals'

Human capital

The labour resources available to households both in terms of quantity, that is the number of members engaged in income-earning activities, and quality, referring to their education, skills and health status.

Social capital

The social resources on which people draw in pursuit of livelihoods, for example social networks and relationships based on reciprocity.

Physical capital

The basic infrastructure, equipment and tools of their trade that people need to pursue a livelihood.

Financial capital

The financial resources people rely on to pursue various livelihood options, such as savings, credit, remittances and transfers.

Natural capital

The natural resources used in the pursuit of livelihoods, for example, land, water and common pool resources.

Source: Carney (1998)

into government institutions and global production processes that rely on the perpetuation of informally contracted labour (Meagher 1995).

To address structural inequities an important policy response is to promote social protection, whether through the provision of social safety nets, basic infrastructure and social services, or enhanced employment and income-generating opportunities. However, in the long run sustainable livelihoods depend upon people having the capabilities to provide for themselves. It may be possible for governments and donors to raise living standards in the short run through coordinated efforts such as the MDGs, but the eradication of poverty ultimately hinges on a buoyant economy, human development strategies, and transformative social protection that see people sufficiently educated and skilled, healthy, socially engaged and politically empowered to provide for their own needs and wants.

This can be illustrated clearly with reference to housing policy. The right to housing is difficult to realise because access to land and housing is tied up with property rights. Even if the urban poor are untiring agents in the quest for shelter, they face structural barriers in land and property markets related to social and resource differentials. Moreover, it has been estimated that for the period 2000 to 2010 the annual need for housing in urban areas of developing countries alone is estimated at around 35 million units, so that 95,000 units a day need to be constructed to meet demand. This is beyond the capacity of most governments, explaining the carefully worded response to the call for housing rights at the Second United Nations Conference on Human Settlements held in Istanbul in 1996 and in the Habitat Agenda that followed (UNCHS 2001a):

> The Habitat Agenda, particularly in its para. 61, clarifies actions and commitments of governments and other stakeholders in order to promote, protect and ensure the full and progressive realization of the right to adequate housing. In this context, it is important to clarify that the 'housing rights framework' does not in any way mean or imply an obligation for governments to immediately provide free housing to all their citizens and/or residents. Yet, as is clearly articulated in the Habitat Agenda, governments are responsible for establishing and facilitating an enabling environment where the potential and capacity of individual households and all other stakeholders in the housing development process are supported.
>
> (UNCHS 2001a)

This policy approach now informs most national housing policies. It recognises that housing is an economic as well as a social good and is cognisant of the limited resources available to governments in low- and middle-income countries. In the following section we explore successive approaches to delivering housing as a critical asset that allows people to provide for themselves in sustainable ways.

Shelter needs and housing policies

People are best able to pursue urban livelihoods when they have access to an established domestic unit that provides adequate shelter and security of tenure. However, ensuring adequate shelter and security of tenure has eluded urban policy makers for decades. One reason lies in the scale of the problem. According to the 2001 *State of the World's Cities* report, 'worldwide, 18 per cent of all urban housing units (some 125 million units) are non-permanent structures, and 25 per cent (175 million) units do not conform to building regulations' (UNCHS 2001b: 30). Even when governments are successful in increasing the number of available houses in real terms, proportionally this can have little effect if the size of the urban population is increasing inexorably. Moreover, even when urban growth is relatively limited, the number of households seeking homes may grow. Attempts to assess housing conditions and housing need are hampered by lack of reliable data, with some countries collecting no statistics on housing at all. For example, some dwellings are excluded from official statistics because they do not conform to official criteria of what constitutes a house, even though people are living there (see Plate 4).

Another challenge in developing effective housing policy relates to the nature of housing itself and the tensions between housing as a shelter need, or *consumption* good, and housing as an *economic* good – an asset

Plate 4 A riverside slum in Mumbai

Source: Philipp Rhode

with a market value. Housing can form an important part of a city's fixed capital, with a buoyant housing market helping to fuel economic growth in other areas. At the individual or household level, housing can be critical to avoiding impoverishment and improving fortunes. While for some people shelter represents little more than a roof over their head, for others even a simple dwelling can be their most valuable possession and its value may increase over time. When informal settlements are regularised, land and housing values increase. The same applies when urban services are extended to slums through upgrading schemes. A house can also improve the economic well being of its owner or occupier. For instance, just having an address makes it easier to find employment. A house can be used as collateral for securing credit, as a site for home-based or small-scale enterprise and can be a potential source of income through renting out rooms.

Viewed over the long term, housing has been seen primarily as an economic good. Indeed, government involvement in the provision of housing is a relatively recent development, with the default position for centuries having been that people would meet their shelter needs through individual effort and private resources. However, during the early development decades, leaders of many newly independent states initiated large-scale housing projects. In some measure, this mirrored trends in the post-war welfare states of Europe at the time, and in part it reflected a desire to redress the injustices of colonialism and to wear post-independence development as a badge of national pride. However, these high-profile housing projects, which were often to European standards of construction, were never meant to house the legions of poor people who came to populate cities. This was, after all, the time when Lipton's urban bias theory was gathering momentum. It was believed that rural–urban migration could be prevented by policy interventions (see Chapter 1) and, related to this, slum clearance was common. When it became clear that the provision of low-cost housing was an imperative, it was assumed that affordable public housing could be delivered incrementally by the state. Typical of the housing provided at this time were four-storey walk-up apartment blocks – now often tenements – which can still be found in cities across Africa and Asia, from Cape Town to Cairo, Kolkata to Cochin.

In the end though, this response was both inadequate to need and unaffordable for most, and construction of conventional housing of this sort never constituted a very large proportion of shelter options in cities of low- and middle-income countries. The failure of governments to directly provide sufficient housing for their urban populations is starkly illustrated in Table 4.1. This shows the performance of Tanzania's National Housing Corporation (NHC) charged with constructing low-income housing in Tanzania's cities. An agency of an avowedly socialist government, Tanzania's NHC fell victim to some of the classic mistakes identified by critics of direct housing provision by government. As Table 4.1 demonstrates, the number of NHC employees mushroomed while productivity (measured as the number of houses built per NHC employee) steadily deteriorated.

Estimates suggest that public housing rarely exceeds 10 per cent of the total housing stock in low- and middle-income countries (UNCHS 1996). Moreover, by the 1960s it was becoming clear that low-income households were finding their own shelter solutions, constructing makeshift dwellings in informal settlements or crowding together in slums (Mangin 1967; Turner 1976). The creative solutions that poor urban dwellers developed autonomously and without the help or interference of bureaucracies was cast in a positive light. Turner (1972) spoke of 'housing as a verb' and argued strongly that if the urban poor were given security of tenure and a plot of decent land they would incrementally achieve for themselves respectable housing. It was argued

Table 4.1 ***The failure of public housing provision in Tanzania***

Year	*Houses constructed*	*Number of employees*	*Houses built per employee*
1962/63	51	10	5.10
1963/64	11	120	0.09
1964/65	1010	250	4.04
1965/66	1220	380	3.21
1966/67	1210	510	2.37
1967/68	1500	660	2.27
1968/69	2400	760	3.16
1969/70	870	860	1.01
1970/71	1340	910	1.47
1971/72	2060	970	2.12
1972/73	1240	1020	1.22
1973/74	1240	1120	1.11
1974/75	310	890	0.35
1975/76	450	1090	0.41
1976/77	150	810	0.19
1977/78	180	970	0.19
1978/79	290	1050	0.28
1979/80	100	1027	0.10

Source: Lugalla (1995: 61)

that the responses of squatters who engaged in self-help housing solutions were just as rational as those of the middle and upper-income classes. As such, they required facilitation and support through cost-reduction measures and the lowering of standards for housing, not evictions and demolitions.

Champions of self-help housing were bent on showing that the urban poor are the best managers of their own housing solutions, and policy makers were more than ready to accept such arguments. After all, governments needed to find ways to finance low-cost housing in ways that were economically feasible. This signalled a major shift in mainstream approaches to housing policy towards the promotion of reduced standards and low-cost solutions in the face of rapid urbanisation. This generally took one of two forms. The first is sites-and-services schemes whereby governments make serviced land available, divide it into plots and sell it to people who are responsible for constructing their own houses. In well-serviced schemes plots will have access to roads, drains, water supply and electricity. Minimalist approaches can include nothing more than a marked out plot and a pit latrine, as with the infamous 'toilets in the veld' to be found on the outskirts of towns across South Africa and far from city centres where residents earn their livelihoods.

In an extension of the sites-and-services idea, sometimes core housing is provided. This comprises a simple structure on a serviced plot, with the idea that the owner can in time build on it. This was a part of the strategy pursued by the post-apartheid South African government, which sought to redress the housing deficit among historically disadvantaged populations, aiming and largely succeeding in building 300,000 houses a year with a minimum of 1 million low-cost houses to be constructed within five years through the vehicle of government subsidies to low-income citizens. However, often the plots are too small for decent size extensions to core housing, as was the case in many instances in post-apartheid South Africa. More generally, the core housing option is often beyond the reach of most poor people. Core housing either ends up in the hands of middle-income people who rent it out, or the people at whom it is aimed end up selling it on because they cannot afford rates and service charges. Even when sites-and-services schemes are deliberately targeted, they can still be exclusionary. For example, in Tanzania the World Bank financed a sites-and-services project requiring applicants to demonstrate they had amassed a certain level of savings, and this provided an almost insurmountable barrier for the very poor (Choguill 1995: 408). Moreover, even minimalist sites-and-services schemes end up marginalising the poorest households in other ways; for example, people do not always have the skills to construct their own dwellings to a standard that meets government regulations and nor can they afford to employ others to build their homes. As a result they risk living in illegal constructions. Women-headed households often lack the ability to construct even the most basic of structures or extensions and are doubly excluded by these schemes as a result (Moser and Peake 1987).

The other main approach to low-cost housing is slum upgrading, which since the mid-1980s has remained the housing policy of choice because it offers a number of important advantages (see Box 4.6). Upgrading schemes are usually designed for areas of a city that are already built up and entail improving existing slums or informal settlements with the inhabitants in situ, rather than attempting to relocate them. Projects usually include improving infrastructure and services such as water supply and sewerage, building health clinics and schools, as well as providing financial services and building assistance to individuals. Sometimes upgrading also includes the rearrangement of at least parts of the layout of a neighbourhood and the extension of security of tenure through legislation. Valued by residents, upgrading is often resisted by urban elites, especially those living adjacent to low income communities

Box 4.6

The advantages of upgrading

- Upgrading preserves existing economic systems and opportunities for the urban poor including proximity to formal sector jobs and employment and entrepreneurial activities in the informal economy.
- It preserves the low-cost housing stock already in existence at its present location, important because even in the case of informal housing and slum settlements, it is easier to destroy than to build, to engage in demolition than reconstruction.
- It preserves and protects the community structures that have built up, such as community organisations and systems of representation, as well as family and neighbourhood level networks of social support that have been established.
- It avoids the socially disruptive alternative of resettlement, which in addition to involving high community costs holds the potential of fuelling discontent and political opposition.

Source: Martin (1983)

slated for upgrading. Opponents to upgrading projects often argue that regularisation legitimises illegal settlements; that providing slums and informal settlements with services raises the value of assets that people have acquired by breaking the law through squatting; and that upgrading gives these settlements permanence, potentially lowering the value of their own homes. Nevertheless, upgrading constitutes the backbone of enabling approaches as evidenced, for example, by the World Bank's 'cities without slums' initiative (World Bank 2001).

The second overall shift in housing policy was a trend that saw governments moving from a provider role to that of 'enabler'. This umbrella approach informed both sites-and-services and upgrading schemes. Here the state provides the legislative, institutional and financial framework 'whereby entrepreneurship in the private sector, communities, and individuals can effectively develop the urban housing sector' (Pugh 1995: 67). Initially the contributions of governments, markets and individuals to the development of housing markets were assumed to be separate and divisible. However, it soon became clear that coordination and cooperation were necessary. For example, governments had to make the necessary reforms to property rights and legislation to ensure that sufficient land was available for development. Private developers engaging in house construction and settlement upgrading required the support of and sometimes co-financing by government, especially when projects included cross-subsidy to poor settlements. Finally, given that individuals or households were expected to pay in order to benefit from housing processes, they required access to credit and this required both public and private housing finance institutions. For its urban strategy paper for the 1990s, the World Bank added to its enabling approach a focus not only on affordability but also on the principles of cost recovery and demand (World Bank 1991). Indeed, as Choguill (1995: 406) has pointed out: 'full cost recovery became a central principle in the design of shelter projects for the urban poor as this is one of the golden tenets of the World Bank for any of its lending projects'.

In order to ensure the healthy functioning of housing markets, on the supply side it was necessary to advocate for a rationalisation of legal and regulatory frameworks, alongside measures to increase competition and reduce monopolies in the construction industry. On the demand side, there was a focus on providing a range of housing options for different income groups rather than concentrating only on housing the urban poor. A critical accompaniment to this was improving access to private housing finance. Problems that soon emerged were that developers were less interested in low cost as opposed to conventional and better-off housing developments. As a result formal housing finance institutions excluded the poorest households. Indeed poorer countries did not always have the necessary housing finance systems to allow for a user-pays principle and a market in housing to develop. Following its strategic housing review in 1992, therefore, the World Bank advocated 'the growth and development of the entire housing sector in its urban and national context' (Pugh 1995: 67). Here the functions of the state continued to be defined by enablement, for example providing tenure rights to slum dwellers and coordinating the provision of infrastructure and services; functions that were reinforced by UN-Habitat through its support to national housing strategies.

Governments have paid little attention to rental housing, which until recently has also been largely ignored by international development agencies as well. However, informal rental markets are important both to low-income landlords and their tenants. Many studies have noted how low-income landlords rent out rooms, backyard shacks and even floor space to even poorer tenants (Beall *et al.* 2002; Crankshaw *et al.* 2000; Gilbert and Crankshaw 1999; Gilbert and Varley 1990; Kumar 1996, 2001). When policy attention is paid to the rental housing option, landlords are frequently portrayed as exploitative. In fact, they are often poor themselves and earn only a modest income from their housing asset (Kumar 2002). Instead of promoting rental solutions, government strategies such as de-densification of overcrowded areas can undermine them. This in turn hampers the drive to eradicate housing poverty while at the same time compromising an important livelihood strategy of many low-income urban dwellers.

In recent years land titling has moved to the fore of the urban development agenda. This policy shift follows de Soto's argument that lack of legal tenure in the form of a title constrains the ability of the poor to realise the value of their assets and discourages self-investment in the improvement of dwellings (de Soto 2001). As a result titling schemes have been initiated throughout the world. While de Soto's vision is compelling that 'the poor are not the problem but the solution' (de Soto 2001: 241), the application of his argument can be problematic in practice. Experience with titling programmes has illuminated pitfalls that may turn a well-intentioned effort to empower the poor and boost the economy into an opportunity for exclusion and graft, resulting in an undesirable allocation of resources. In a survey of urban tenure and property rights issues, Geoffrey Payne observes that land speculation in anticipation of a regularisation scheme based on titling may lead to price inflation that effectively excludes the poor from accessing land or rental housing through the market (Payne 1997: 46). He goes on to suggest that such schemes may encourage the development of new extra-legal settlements whose residents anticipate being granted titles to the land they occupy. Furthermore, the expected benefits of formal titling may not materialise if a well-regulated financial system is not already in place and able to provide credit for land or housing improvements (Payne 1997). The titling process may also provide an opportunity for well-connected or well-educated individuals to seize land, having better access to the institutions and organisations managing the titling process.

There is no doubt that effectively addressing housing poverty requires the continued and active participation of governments because the task cannot be left either to the vagaries of the market or the tenacity of poor people. An important role for the state is to guarantee security of tenure. Formal individual titling is expensive and difficult, sometimes beyond the immediate capacity of the cadastral services of countries and cities. However, the protection of tenure as a fundamental right need not depend entirely on formal property titles, although these are desirable. Other, sometimes more informal, forms of tenure can help address the basic needs of the poor and improve the functioning of land, housing and rental markets in low income areas. For example, governments can stop engaging in forced evictions, keep records of land use through incremental upgrading programmes and promote occupancy rights if not full land titling (UN-Habitat 2003b: 171). However, the problem is often not a lack of possible solutions, but a lack of political will – a subject we will return to in Chapter 7. Where the political will does exist, the results can be truly remarkable (see Box 4.7).

Box 4.7

Sri Lanka's Million Houses Programme

Sri Lanka managed to make a significant dent in its shortage of adequate housing despite ongoing conflict between Tamil and Sinhalese groups. It has achieved this through a creative programme combining government initiative with market forces, taking advantage of a long history of participatory decision-making at local government levels.

Up until the mid-1970s Sri Lankan housing policy was typical of countries in low- and middle-income countries: limited public housing was provided in the form of rental flats, and subsidised interest rates for borrowing targeted mainly at middle-income households were provided through a national housing fund. Residents of informal settlements fell outside the scope of policy except for periodic evictions until the Aided Self-Help (ASH) programme was established, providing for community participation and low-cost self-help assistance with loan finance from government. A review of housing policy in the early 1980s found a severe housing deficit and the government launched its Million Houses Programme (MHP) between 1984 and 1989.

The MHP expanded the self-help programme, offering 19 choices on the size and use of loans for new or upgraded dwellings – 16 of which were targeted at low-income households. Loans were available at concessional rates and with a 15-year repayment period. The programme fulfilled its targets, with 50 per cent of the loans being used for upgrading, 31 per cent for new extensions and latrines and 15 per cent for completely new housing.

There are problems with replicating the MHP's success elsewhere given the context-specific factors that made it successful. Furthermore, Sri Lanka was able to finance the programme (and its successor, the 1.5 Million Houses Programme between 1990 and 1995) only because of extensive financial support from international donor agencies. Nevertheless, the programme developed the national housing sector, added to the economic strength of the building and building materials industry, and led to a decrease in housing poverty, with better quality housing and lower occupancy rates (i.e. more rooms per capita) for many more people. As such, Sri Lanka's experience offers important lessons for designing and operating mass low-income housing programmes.

Source: Pugh (1995)

Conclusion

In order to effectively analyse, measure and address urban poverty and vulnerability it is necessary to go beyond aggregate statistical assessments and examine the complex of vulnerabilities confronted by the urban poor. Given the monetised nature of urban economies, securing basic necessities requires a steady income. But cash alone is not enough to ensure sustainable livelihoods, which depend upon the expansion of people's capabilities. Achieving this involves investments in intangible assets such as skills, education and social networks as well as material assets such as housing. Livelihoods and asset accumulation are best pursued from the security of adequate shelter where tenure is assured. Not only does this provide a form of social protection, but also housing can provide the respectability and dignity of an address and serve as an economic asset.

Policy trends increasingly see the pursuit of livelihoods and shelter as the responsibility of urban dwellers, facilitated and enabled by governments. Nevertheless it is important that urban livelihoods and housing be understood and addressed not only at the level of individuals, households and communities, but also at the level of the city as well. As we will see in Chapter 5, the effective management of the urban built environment is essential for both poverty reduction and the efficient functioning of urban economies. And in the long run, the cultivation of a dynamic city economy is essential for ensuring increasing opportunities for the urban poor.

Summary

- **Measurements of *absolute poverty*, such as poverty headcounts based on income thresholds, tell us little about relative deprivation within particular countries or the multidimensional nature of poverty.**
- **Consequently there has been increasing interest in ideas of *capability deprivation*, which take into account non-monetary and relative aspects of poverty and vulnerability.**
- **Many assessments of poverty ignore its spatial dimensions, such as the difference in character between rural and urban poverty. While the former is still more widespread in absolute terms, the latter is on the rise and is characterised by different processes.**
- **The urban poor are vulnerable due to such factors as their reliance on a monetised economy, high-risk informal working environments, particular urban disease vectors, crime, and housing located in areas at risk of flooding, fire and eviction or bulldozing.**
- **In the wake of structural adjustment various social safety nets were put in place for the poor, often through social funds. These generally proved inadequate and unsustainable. More recently a focus on *sustainable livelihoods* has come to the fore.**
- **Concepts such as sustainable livelihoods and *asset accumulation* have proved very useful in developing broader approaches to poverty reduction, though they do not directly address fundamental structural inequalities.**
- **The question of who should provide housing for the poor, and how, illustrates many of the problems associated with alleviating urban poverty. Governments have failed to keep pace with urban shelter needs but their role must remain central.**

- **The two main approaches to tackling the problem have been 'sites-and-services' schemes and 'slum upgrading'. Both face considerable practical and political problems, and providing security of tenure for the urban poor remains an enduring challenge.**

Discussion questions

1. Discuss the problems involved in measuring and assessing poverty, with reference to the terms 'absolute poverty', 'relative poverty' and 'capability deprivation'.
2. What can we gain by making a distinction between rural and urban poverty?
3. How have approaches to tackling urban poverty changed in response to dominant trends in development thinking since the 1970s?
4. Why might it be helpful to think of poverty in relation to livelihoods and assets rather than monetary income?
5. What role should governments play in the provision of housing for the urban poor?

Further reading

Aldrich, Brian and Ranvinder Sandhu (eds) (1995) *Housing the Urban Poor: Policy and Practice in Developing Countries*, London: Zed.

Devas, N., with P. Amis, J. Beall, U. Grant, D. Mitlin, F. Nunan and C. Rakodi (2004) *Urban Governance, Voice and Poverty in the Developing World*, London: Earthscan.

Kumar, Sunil (1996) 'Landlordism in Third World urban low-income settlements: a case for further research', *Urban Studies*, 33(4–5): 753–782.

Moser, Caroline (ed.) (2007) *Reducing Global Poverty: The Case for Asset Accumulation*, Washington, DC: Brookings Institution, pp. 83–103.

Rakodi, Carole with Tony Lloyd-Jones (eds) (2002) *Urban Livelihoods: A People-Centred Approach to Reducing Poverty*, London: Earthscan.

Ravallion, Martin, Shaohua Chen and Prem Sangraula (2007) *New Evidence on the Urbanization of Global Poverty*, Policy Research Working Paper 4199, Washington, DC: World Bank, available at http//econ.worldbank.org/docsearch (accessed 9 June 2008).

UN-Habitat (2005) *Financing Urban Shelter: Global Report on Human Settlements 2005*, London: Earthscan.

Useful websites

Eldis urban poverty page: www.eldis.org/go/topics/resource-guides/poverty/urban-poverty
Livelihoods Connect: Support for Sustainable Livelihoods: www.livelihoods.org
UN-Habitat Housing Rights Programme: www.unhabitat.org/categories.asp?catid=282
World Bank urban poverty page: www.worldbank.org/urban/poverty

5 Managing the urban environment

Introduction

In Chapter 4 we showed how poverty and vulnerability are manifest in cities, illustrating in particular how tenuous livelihoods and inadequate housing increase social disadvantage and risk. Here we take forward our concern with vulnerability by focusing on the risks associated with deficient urban services and neglect of the urban environment. Inadequate water supply and sanitation, poorly managed transport systems, as well as unanticipated disasters all increase risk and urban inequality. We consider these issues in terms of their impact on urban residents, on the overall functioning of cities, as well as how the urban environment contributes to broader environmental concerns, including climate change. Many such problems relate to bad urban management. The potential to equitably and efficiently manage urban services exists, although this is not always realised. We are equally capable of managing wider environmental hazards and risks, such as atmospheric pollution, yet here too we often fall short. When we fail we threaten the capabilities of the most vulnerable urban dwellers, the optimum functioning of us all and, by putting at risk the sustainability of our environment, the freedom of future generations.

In this chapter we proceed with an examination of essential urban services – water supply, sanitation and the removal and disposal of solid waste or garbage – as well as the impact on urban residents of inadequate provision. Next we consider changes in approach to how cities are run, looking at the shift from public administration and service delivery to urban management, and show how this shift mirrors broader changes in development policy. In particular we reflect on how the introduction of new public management, which accompanied the ascendancy of neo-liberal development orthodoxy in the 1980s and 1990s, infused urban management discourse and impacted on access to essential urban services. We then focus on transport and atmospheric pollution and how poor management of the urban environment can affect the global environment. The chapter concludes with a discussion of sustainable development and cities, exploring the two-way and sometimes contradictory relationship between the urban and global environments with particular reference to climate change.

Critical urban services: water, sanitation and solid waste management

The health and well-being of urban residents everywhere depends upon the effective, efficient, and affordable provision of safe water supply, sanitation and the removal and disposal of solid waste or garbage. These constitute the most critical urban services upon which managing the disease environment of a city depends. Yet globally, 1.1 billion people lack safe water and 2.4 billion have no access to adequate sanitation, with over 90 per cent of them living in Africa and Asia (WHO/UNICEF 2000) (see Table 5.1). Poor delivery of such services has particularly pernicious effects in cities where population density and residential concentration multiply the risk of disease transmission. For example, water-related infections are transmitted by the ingestion of materials contaminated by human or animal excreta or other vectors, including cholera, malaria,

Table 5.1 ***Urban water and sanitation provision by region***

Region	*House or yard connection for water (%)*	*Connected to sewer (%)*
Africa	43	18
Asia	77	45
Latin America and the Caribbean	77	35
Oceania	73	15
Europe	96	92
North America	100	96

Source: WHO/UNICEF (2000); these figures are based on information provided for 116 cities

typhoid, dengue, infectious hepatitis, diarrhoeal diseases and dysenteries. Such diseases are among the most significant causes of illness and death in low- and middle-income countries, affecting 900 million people a year and leading to the deaths of 2 million children annually (Castro 2005: 751). Moreover their impact is not equally shared. It is women and children who suffer most, and child mortality among the urban poor is often as high as for rural children – and sometimes even higher (Tannerfeldt and Ljung 2006: 64). The prevalence in cities of diseases such diarrhoea, tuberculosis and cholera is closely related to unequal access to services for the urban poor. Indeed, there is strong evidence that 'health outcomes in the urban environment derive ultimately from the socio-economic more than the physical environment' and that poverty 'remains the most significant predictor of urban morbidity and mortality' (Bradley *et al*. 1992: viii).

The true magnitude of the crisis of water supply and sanitation in cities of low- and middle-income countries is difficult to assess as statistics on urban services are patchy and frequently unreliable. Most census and household surveys used to collect data on urban services do not ask the critical questions: how many people share facilities (in some cases thousands of people share one latrine), how long does it actually take to access facilities (sometimes women queue for hours at community standpipes), and how regular and safe is the supply of water (in some cases water is filthy and available for only a couple of hours a day). Hardoy *et al*. (2001: 65) point out that 'it is certain that tens of millions (and perhaps hundreds of millions) of urban dwellers classified in official statistics as having "safe water" still face great difficulties in obtaining clean water and sufficient water for good health'. There are also problems of definition when measuring the quality of service provision. The joint WHO and UNICEF *Global Water Supply and Sanitation Assessment 2000 Report* (WHO/UNICEF 2000) uses the category of 'improved' water supply and sanitation to describe better levels of access; however, the criteria for 'improved' are not uniform, nor are they synonymous with adequate and safe provision, as many have noted since (Bartlett 2003; UN-Habitat, 2003a). For example, while the number of African urban dwellers without 'improved' provision for water is estimated at 44 million (or 15 per cent of the population), the number without adequate provision is estimated at 100–150 million (or between 35 and 40 per cent of the population) (UN-Habitat 2003a).

The number of people without adequate sanitation is thought to be even higher (see Plate 5). For example, estimates suggest that between 50 and

Plate 5 Domestic sewage flowing into city streets in São Paulo

Source: Lucy Earle

60 per cent of Africa's urban population does not have access to adequate sanitation (UN-Habitat 2003a). Indeed, levels of sanitation provision are far worse than is often supposed. In cities such as Addis Ababa, Dar-es-Salaam, Brazzaville, Kigali, Kinshasa, Harare and Lagos, more than 90 per cent of the population live in homes with no connection to a sewer, and overall it is 'unlikely that more than one-third of sub-Saharan Africa's urban population has access to sanitation that is adequate in terms of convenience and the safe disposal of human excreta' (UN-Habitat 2003a: 31). The consequences can be unpleasant and present serious health hazards. In the absence of adequate sanitation facilities in densely populated areas of Nairobi, slum residents rely on 'flying toilets'. This involves putting human waste in a plastic bag and throwing this 'scud missile' on to the nearest roof or pathway (M. Davis 2006: 136). In one low-income area it is estimated that 80 per cent of residents dispose of their excreta in this manner (UN-Habitat 2003a: 27). As Mike Davis (2006: 139) has put it: 'The global sanitation crisis defies hyperbole.'

The situation in South Asia is only marginally better. Accounts from Dharavi in Mumbai, often referred to as Asia's largest slum, show women to be affected disproportionately by the lack of adequate sanitation facilities. Many are forced to defecate in the open and stigma prevents them from going in daylight hours so they have to wait until after nightfall or go before sunrise. One woman from a low-income area of the city told of how they had to use the railway tracks:

> There were public lavatories, but they were some distance away – about half an hour walk. They used to be so dirty that we did not feel like using them. And there were such long queues! Instead of using those filthy lavatories, we used to go on the tracks after ten at night or early in the morning at four or five o'clock.
>
> (Sharma 2004: 22)

The initiative to build community-run toilet blocks unsurprisingly often comes from women. In the Ganesh Murthy Nagar slum in Mumabi, there was until recently 'one small, smelly toilet for a population of 10,000' (WHO 2002). Local women in the slum who suffered the most from poor sanitation provision now manage new toilet blocks as part of a scheme supported by the World Bank. The building of new community-run toilets is taking place in many of Mumbai's slums, whose residents, it is said, 'are ushering in a quiet sanitation revolution' (WHO 2002).

Solid waste management (SWM), or the collection and disposal of garbage, constitutes the third critical urban service. SWM is often viewed as the Cinderella of urban services: neglected in favour of water supply, which is vital for life, and sanitation services that are more evidently linked to issues of health and dignity. Nevertheless, poor SWM can impact negatively on water supply and sanitation, resulting in blocked drains and sewers. This in turn can lead to the build-up of domestic and human waste, polluting water sources and breeding disease. Most large cities in low- and middle-income countries have inadequate waste collection and disposal services. In Dar Es Salaam, for example, with a population expanding at 7 per cent a year and where 75 per cent of the population live in unplanned areas, only 10 per cent of the 2000 tons of solid waste generated daily is actually collected (Kironde 1999). A similar situation is found in Lusaka, where around 90 per cent of the 1400 tonnes of municipal waste produced every day is left uncollected (Hardoy *et al.* 2001: 80). Where collection does take place in these cities, it is invariably in commercial centres and wealthy districts. While residents of high income areas are able to access regular municipal waste collection services through their political influence or by paying for private collection services, this is not an option for residents of low-income areas. Here people are forced to burn rubbish or dump it in ditches, lakes, rivers, on pavements and by the roadside. The potential for the pollution of water sources is enormous – in São Paolo, half of the city's favelas are located on the banks of reservoirs that supply the city and yet it is here that informal settlement dwellers are destined to throw their waste (Taschner 1995: 193).

Areas not adequately serviced by official waste collection services rely instead on community provision, often executed through schemes employing or contracting waste pickers who retrieve recyclable material throughout the city. In Bangalore, the centre of India's information technology industry, the collection and disposal of garbage has not kept up with waste generation as the city becomes increasingly affluent. Once known as India's 'Garden City', Bangalore is now often dubbed 'Garbage City' and even better-off neighbourhoods form community groups to address inadequate waste collection in their areas. Some, like Waste Wise, have grown to become city-wide organisations, providing advice and assistance to other neighbourhood organisations wanting to engage in community based waste collection by procuring the services of informal waste pickers who work as refuse collectors and street cleaners (Beall 1997a: 952). In poorer settlements, community involvement in waste management is more likely to take the form of self-help initiatives. There are many examples of such schemes across Africa, Asia and Latin America and sometimes efforts are made to integrate them into official waste management systems. This has been done very successfully in Manizales, Colombia for example, where *recicladores* or 'scavenging' communities are now formally incorporated into the running of the city's SWM system.

The ultimate disposal of waste constitutes a real problem for many cities, even where collection is taken care of satisfactorily. Managed landfill sites are rare in low- and middle-income countries. Instead, rubbish ends up in open landfills that are often tens of metres high and in serious danger of collapse. Many are located near informal settlements; in some cases people who make their livelihoods from retrieving and recycling waste materials deliberately form residential communities on or near city dumps. Most notorious was 'Smokey Mountain' in Manila, Philippines, which was closed down in 1995 because garbage rotted at temperatures so high that it combusted, leading to fires and parts of the gigantic waste heap collapsing, killing many people. As dump sites become inadequate for the volume of waste a city generates or when poor collection prevails, waste encroaches further and further into the living spaces of the urban poor. The problem of uncollected waste can have extremely far-reaching implications in concentric circles of primary, secondary and tertiary environmental effects. As such, poor SWM can affect whole urban populations as well as the efficiency and reputation of a city.

A stark illustration of inequality in cities is the fact that higher volumes of garbage are associated with rising levels of affluence, cheaper consumer products, built-in obsolescence, increased packaging and the demand for convenience products. The content of waste has changed concomitantly, with kitchen and organic waste giving way to higher proportions of recyclables such as paper, cardboard, plastics and metals. Accompanying rising affluence are levels of poverty that see legions of men, women and children who make a living from this waste. These are the scavengers who have become the leitmotif of urban poverty, living on the garbage dumps of Manila and Mexico City and rummaging through the bins and waste heaps of Kolkata and Karachi (see Plate 3, p. 120). They are the most visible symbols of the paradoxical relationship between the environment and inequality: namely that affluence produces abundant waste, while poverty does not; that poverty encourages efficient reuse and recycling of waste materials, while affluence does not; and that urban livelihoods built on resource conservation and recycling, ironically and tragically, are predicated upon persistent inequalities in income and consumption (Beall 1997b).

For all critical urban services it is the poorest districts most in need of serious assistance that pose the greatest risks for those engaged in service provision. When local governments lack the resources or political will to invest in provision, private actors often fill the gap, naturally giving preference to wealthier neighbourhoods. In better-off areas the terrain is usually more suitable for the installation of bulk infrastructure, the locations are more central and accessible, and the payment of user charges is more predictable. In slum areas, on the other hand, numerous risks abound. Informal or illegal settlements are often located far away from existing trunk infrastructure such as water mains and sewerage networks, where the costs of upgrading infrastructure and improving service provision are high because the terrain is inaccessible or dangerous and land tenure is uncertain or illegal, and where layouts are dense and hence difficult to access. Even when policy is geared actively towards extending water and sanitation systems to poorer areas, historical patterns of urban growth and development can render this difficult (see Box 5.1). There are also issues of political will. For example, the new Johannesburg Metropolitan Council faced significant hurdles in the post-apartheid era when trying to extend services to previously underserved areas, while at

Box 5.1

The crisis of infrastructure in Lagos

Lagos is the largest city in Sub-Saharan Africa with a population set to reach 17 million by 2015. Unequal access to urban services dates back to colonial times when the segregation of exclusive enclaves was encouraged and investment in infrastructure was geared towards wealthy areas. Indeed, at times of water shortage, colonial authorities instructed householders not to let their staff bathe and during the 1920s the inadequate reach of water and sanitation infrastructure led to outbreaks of bubonic plague in poor areas. In later decades unregulated urbanisation led to a deepening of inequality within the city, which expanded haphazardly with few concerted attempts to manage this growth. At independence there was reportedly only one skilled engineer in charge of the city's entire water distribution system. Only 10 per cent of dwellings were connected to the municipal water system, while inequality in terms of sewerage connections provided a striking disjuncture between a 'showcase modernity' – reflected in the construction of prestige projects such as the National Theatre – and the city's continuing inability to provide basic infrastructure. The economic impact was dire with most firms having to spend over 20 per cent of their earnings on providing their own sources of water, electricity and other basic services.

The discovery of oil in the Niger delta has only worsened the situation in the city. After the failure to implement the 1980 Lagos Master Plan, few attempts have been made to deal with the city's problems in a strategic way. Basic services have degenerated yet further with less than 5 per cent of households now connected to piped water and less than 1 per cent linked to a closed sewer system – this proportion largely consisting of hotels and rich compounds. Other forces are at work perpetuating unequal access to water, including hostility and violence from water vendors who benefit from unequal access by selling water to the poor and who resist municipal attempts to extend water supply. Private investors are only willing to invest in the richer areas and are averse to taking the risks associated with installing basic infrastructure in conditions of such extreme poverty, thereby perpetuating already deeply entrenched divisions.

Source: Gandy (2006)

the same time maintaining service levels in better-off suburbs used to high standards of provision. Politically, the Metro Council could neither neglect the economically powerful and well-organised ratepayers in wealthy neighbourhoods, nor ignore those clamouring for the services from which they had formerly been deprived (Beall *et al.* 2000: 850).

The challenge of equitable provision is complicated by the question of appropriate standards. There is a danger in setting the bar too high:

if piped water and access to waterborne sewerage for every house is considered paramount, this can result in incremental progress that inevitably begins with the areas where installing such infrastructure is most feasible and practical – almost always in and near wealthier neighbourhoods where services already exist – thus perpetuating inequalities. UN-Habitat argues that 'it is better to provide a whole city's population with safe water supplies by means of taps within 50 metres of their home than to provide only the richest 20 per cent with piped water to their homes' (UN-Habitat 2003a: 3). At the same time, there are normative questions about whether it is acceptable in middle-income countries to deliberately provide a lower standard of service to people living in certain areas. These questions become politically charged when people living in those areas are predominantly of one religious, ethnic or racial group. Patrick Bond (2000) argues in the case of South Africa that a country classified by the World Bank as 'upper-middle-income' should not adopt excessively low standards of service provision for poor areas on the grounds that it is not only inequitable but also inefficient. This is because it is extremely difficult, after installing yard taps rather than in-house reticulated water, unpaved rather than paved roads, and 5–8 amp rather than 20 or 60 amp electricity, 'to incrementally upgrade infrastructure – particularly sanitation systems – from pit latrines to waterborne sewage, resulting in permanently segregated low-income ghettoes' (Bond 2000: 54). Hence solutions based on lowered standards may seem 'appropriate' in the immediate term but can have undesirable consequences in the long run.

Striking the right balance between meeting immediate needs and serving long-term development objectives is difficult, although the long-term benefits of investment in critical urban infrastructure and services are undeniable. And the impacts of better facilities reach far beyond improvements in health, having positive knock-on effects on household incomes, school attendance, family and community relations and many other factors (WaterAid 2001: 2). In a study in Addis Ababa it was found that, among other benefits, improved water supply meant the time spent collecting water was reduced from 6–8 hours to 5–20 minutes, resulting in much less fatigue for women. The incidence of eye diseases and postnatal infections was also reduced, alongside improved menstrual hygiene, increased self-esteem and more time available for family and community activities (WaterAid 2001). Thus the overall environmental and social impact of improving urban water and sanitation is much more far-reaching than is often supposed. A further example is provided in the case of Mumbai, where around half of Mumbai's 20 million residents live in slums. It is women and girls who queue in order to get just a fraction of their family's daily water requirements (Sharma 2000). In some areas tankers deliver water and female members of the household are expected to wait each day for the delivery. This can have far-reaching implications – as one Mumbai slum dweller explained: 'My sister stays at home and waits for the tanker, so she has stopped going to school' (InfoChange News 2005).

The case for improving essential urban services is incontrovertible, both for reasons of equity, to improve the health and capabilities of the urban poor, as well as to ensure the efficient functioning of cities and urban populations more generally. Given that this is the case, it is difficult to understand why there are such gaping deficits in their provision. Part of the explanation lies with the challenges of urban growth, discussed in Chapters 1 and 2. Chapters 3 and 4 showed further that urban economies are not growing in ways that ensure benefits extend to a majority of urban citizens. This too places enormous challenges on city governments trying to meet the demands of delivery to growing numbers of poor people who find it difficult to pay for services. All these are important factors but a critical element remains when seeking explanations for infrastructure and service deficits: weak urban management.

Managing urban services

In Chapter 1 we discussed how states were seen as the primary agents of development in the immediate post-war period. During the 1980s state-led development gave way to a global orthodoxy based on the view that markets trump states as drivers of development. The international political climate at the time was preoccupied with the failure of socialism in the former Soviet Union and disillusionment with central planning. The merits of market-driven approaches were therefore seen as self-evident. Central to the resulting global policy orthodoxy were efforts to reduce the role of governments in service delivery and to open up markets to foreign and domestic private sector investment. Some countries, especially in East Asia, did not abandon the developmental state and their continued success gave rise to efforts from the late 1990s to 'bring the state back in' and to pay greater attention to institutional reform and governance. All these trends played themselves out in how people thought about and engaged in urban management.

Under state-led development the administration of public goods and services was conducted according to bureaucratic principles. The assumption was that city administrators would pursue policies in accordance with their own professional expertise and in the public interest. The new orthodoxy questioned this, seeing bureaucrats as self-serving and opportunistic. The causes of poor performance have been extensively commented on in a familiar litany, caricatured as follows:

> [Public] officials and their workers pursue their own private interests rather than those of the public good; government spending and hiring is overextended; clientelist practices are rampant, with workers being hired and fired for reasons of kinship and political loyalty rather than merit; workers are poorly trained, and receive little on-the-job training; and tying it all together, badly conceived programmes and policies create myriad opportunities for graft and other forms of 'rent-seeking'.
>
> (Tendler and Freedheim 1994: 1771)

Rent-seeking has been described as 'a polite word for corruption' (Harvey 1991: 141). It refers to the practice of extracting resources and privileges by virtue of political power or bureaucratic position. For example, government officials might, in addition to their salaries, make private use of office transport and other public resources, or use privileged information obtained in the course of public service to advance private businesses. The market was seen as the appropriate response to solving such problems, the assumption being that competition would substitute for corruption.

These ideas were encapsulated in an administrative paradigm known as new public management (NPM). The term was coined by Christopher Hood (1991) to refer to the organisational changes in public administration designed to align it more closely with the methods and management systems of business. In essence, NPM involves strategies to increase market competition, give greater choice to users (demand-driven approaches) and promote efficiency in service delivery (cost-sharing). In practical terms this has meant 'increasing focus on the customer, user fees or charges, contracting out of service delivery to the private and voluntary sectors, public-private partnerships, and outright privatization' (Batley and Larbi 2004: 49). In time it became clear that the state could not be bypassed completely and so a prerequisite of the NPM was to inject into service delivery a more entrepreneurial style, with government moving from a concern 'to do' towards a concern to ensure that 'things get done' (Batley and Larbi 2004).

The reform of urban management practices was informed by this shift towards NPM and advocated by international agencies. The Urban Management Programme, jointly funded and run by the World Bank, UNDP and UN-Habitat, promoted a 'new urban management' agenda, which was most clearly expressed in the World Bank's publication, *Urban Policy and Economic Development: An Agenda for the 1990s* (World Bank 1991). Seeking to give city managers more autonomy over operational matters and casting the local state in the role of overseer and coordinator, the idea was that essential urban services would be provided through private solutions or public–private partnerships. Privatisation is defined generally as 'the shifting of a function, either in whole or in part, from the public sector to the private sector' (Butler 1991: 17). However, there are various arrangements by which a state can transfer responsibility to private providers. Box 5.2 summarises these in relation to water and sanitation. The push for private solutions has had significant impacts on urban service delivery and on the affordability of services for the urban poor.

Box 5.2

Types of contracts for water and sewerage projects

Service contract

The most basic and short-term arrangement, whereby a private company is contracted to perform a discrete activity such as fixing pipes or collecting bills.

Management contract

A more involved transfer of responsibility where the management of certain operations is handed over to the company but investment and expansion remains with the state.

Concession

The private contractors manage the whole utility at their own commercial risk. They are also required to invest in the maintenance and expansion of the system.

Build-own-transfer (BOT)

Similar to concessions, BOT contracts are usually used for new projects where the private contractor is responsible for constructing infrastructure from scratch.

Divestiture

A model of full privatisation in which a private company purchases the assets from government and takes over their operation and maintenance as a commercial business and on a permanent basis. This has only been adopted so far in England and Wales.

Joint ventures

An arrangement rather than a contract whereby private investors form a company with the public sector, which then takes on a contract for utility management.

Sources: Budds and McGranahan (2003); UN-Habitat (2003a)

The prior neglect of urban services meant they became inevitable targets for reform. But the introduction of private sector involvement in large strategic enterprises such as urban service delivery was inevitably going to be controversial. The case for private provisioning was made primarily on economic but also environmental grounds. First, it was argued that precious commodities such as water have an economic value that should be reflected in prices. Second, when the public sector provides scarce goods such as water at subsidised rates, it was argued with some credence that they are overused. When produced by private operators, and with the full costs of production passed on to consumers, the belief was that incentives are put in place to use such scarce resources more parsimoniously. The argument was also made in terms of equity on the grounds that markets are more likely than states to deliver the greatest good to the most people over time.

Hence, under the new global orthodoxy water came to be viewed not simply as a public good but also as an economic good to be provided according to efficiency principles with equity outcomes in the long run. This position was clearly articulated in the Dublin Principles of the International Conference on Water and the Environment (ICWE) in 1992. Principle number four was that water has an economic value in all its competing uses and should be recognised as an economic good. However, the conference also recognised access to water as a basic right:

> Within this principle, it is vital to recognize first the basic right of all human beings to have access to clean water and sanitation at an affordable price. Past failure to recognize the economic value of water has led to wasteful and environmentally damaging uses of the resource. Managing water as an economic good is an important way of

achieving efficient and equitable use, and of encouraging conservation and protection of water resources.

(ICWE 1992)

There are difficulties, both theoretical and practical, in seeing water as a private economic good. Water, like sanitation, possesses all the characteristics of a natural monopoly: it depends on large investments in a unified system of infrastructure capable of delivering a functional service at a city-wide scale. In addition water is a public good, being necessary for the life and health of all citizens. Together these characteristics suggest that water should be provided by public agencies. However, public service provision, which is invariably subsidised, favours better-off households with piped connections and an existing reliable supply of water. Public provision in many cities has not ensured water connections for low-income people, who often have to rely on purchasing water from informal vendors. In some cities it is possible to find someone without access to a tap paying up to 50 times more for clean water than a resident living a stone's throw away in a fully serviced neighbourhood (see Table 5.2). For those who do have access to community standpipes or reticulated water, this does not necessarily mean the water supply is reliable or safe to drink. In other words, not only do poor people pay more for water, but also it is often for water of lesser quality: 'Put bluntly, the poor pay more for their cholera' (Stephens 1996: 16).

If we look at *how* water is provided, we can see that it can have private good elements. Hence there is a point at which water as a basic right needs to be delinked from water as a private good. This relates in turn to

Table 5.2 ***The cost of accessing water in some Asian cities***

	Cost of water per cubic metre (US $)			*Difference between price of house connection and price of private vendor*
	House connections	*Public tap*	*Water vendor*	
Bandung	0.38	0.26	3.60	847%
Bangkok	0.30	—	28.94	9547%
Chonburi	0.38	—	19.33	4987%
Dhaka	—	0.08	0.84	—
Kathmandu	0.18	0.24	2.61	1350%
Manila	0.29	—	2.15	641%
Mumbai	0.07	0.07	0.50	614%
Port Vila	0.42	0.86	8.77	1988%
Seoul	0.25	14.13	21.32	8428%

Source: UN-Habitat (2003a)

the level of service provided. In other words water becomes an economic good beyond a certain level of consumption. The difficult issue is settling on the point at which this occurs, a highly political decision. Is it after a certain number of kilolitres have been consumed; or when it is provided by way of more than a community standpipe; when a household has a tap in a private yard; when there is indoor plumbing; or when water is used to fill domestic swimming pools and water vast lawns? In cities that are home to dominant elites but that are also struggling to provide the most basic of services to large and growing numbers of poor people, these are difficult questions.

The decision to privatise or not depends on many factors, and therefore it cannot be guided by ideological mantra. The nature of the service to be provided – whether it is a natural monopoly, a public good or an economic good – can greatly influence the organisational arrangements for delivery of that service, with some services lending themselves more naturally to private sector provision than others. Richard Batley (1996) examined a number of urban services with a view to assessing which would be most likely to benefit from private involvement, and what form that involvement should take. Batley (1996) placed sanitation close to the public goods end of the spectrum, arguing that because poor sanitation has huge negative externalities, there is a strong case for keeping it public, except for specific works that can be contracted out to private firms. By contrast, he suggested that solid waste management is an urban service that has both public and private good elements and as such can be separated into a number of discrete services. For instance, he identifies waste *collection* as an ideal area for private provision as here one can exclude those who do not pay for a door-to-door service. However, solid waste *disposal* has the properties of a public good in that it is hard to exclude non-payers from the benefits of waste disposal, while the negative impacts of poorly managed waste dumps and landfill sites affect everyone.

Water privatisation has been the most contentious in part because of its qualities as a natural monopoly and in part because of the aggressive scramble for contracts and concessions that ensued in low- and middle-income countries during the era of neo-liberal reforms. The water sector worldwide is dominated by a small number of multinational corporations, which together control over 80 per cent of the privatised water and sewerage market (UN-Habitat 2003a: 180). French water companies were the first to move into the capital cities of middle-income countries during the 1990s when markets opened up in cities such as

Santiago, Buenos Aires, Manila and Jakarta. British companies such as Thames Water, Severn Trent and BiWater swiftly followed with financial support from the World Bank. Nevertheless, private investment in water infrastructure has fallen steadily since its peak in 1997, a phenomenon accompanied by high-profile cancellations or renegotiations of contracts (Harris 2003). This followed a realisation on the part of lenders and private operators alike that 'the water and sewerage sector is both more complex and less profitable than originally anticipated' (UN-Habitat 2003a: 178).

By the late 1990s, water companies themselves were publicly admitting that they had been ill equipped to meet expectations in cities of low- and middle-income countries (World Development Movement 2005). A BiWater executive who unsuccessfully brokered a concession in Zimbabwe explained that 'from a social point of view these kinds of projects are viable, but unfortunately, from a private sector point of view, they are not' (Joy and Hardstaff 2005: 16). From the perspective of many poor people, the privatisation of urban services has rendered them unaffordable. In some cases, prices for water rose dramatically: 500 per cent in Conakry, Guinea over five years; 200 per cent in Cartegena, Colombia over eight years; an immediate rise of 68 per cent in Cochabamba, Bolivia; and a doubling of water rates in Tucuman Province in Argentina (Joy and Hardstaff 2005: 17). Needless to say, the failure of privatisation to improve services for the majority of urban dwellers and the accompanying price hikes has led to substantial disillusionment (see Box 5.3). It has also led to popular demonstrations, non-payment campaigns, the destruction of water meters and massive protests in cities across the world.

Privatisation failures in the water sector exposed, in extremis, what became increasingly clear for other services as well: that private firms are no more efficient than the public sector in the provision of public services (Estache and Rossi 2002). In general privatisation does not guarantee that the behaviour of managers and staff will be more effectively monitored than in state enterprises (Chang 2003: 208). With regard to urban services, private ownership did not necessarily guarantee more effective services than under public provision (Batley 1996). When improvements were made, it was usually at the expense of universal coverage, with low income areas being excluded because they proved difficult or expensive to reach. Private contractors are unwilling to take the risks involved in investing in the very poorest parts of cities and governments are then forced to take on a compensatory role, coping with the most difficult and

Box 5.3

The failed privatisation of water and sanitation in Buenos Aires

The financial and economic crisis in Argentina in the early 1990s provided an opportunity for the international financial institutions to prescribe widespread privatisation of services as a solution. This was not difficult to justify given the deplorable track record of Obras Sanitarias de la Nacion (OSN), Buenos Aires' water and sanitation authority, and the city was soon poised to become a flagship example of water privatisation. To make OSN more attractive to bidders the government raised service rates several times to the extent that during the last 26 months of OSN management, water and sanitation tariffs paid by customers increased by over 200 per cent.

The concession was awarded to a consortium, Aguas Argentinas SA, with Suez as the leading shareholder and including Aguas de Barcelona, Vivendi, Lyonnaise-des-Eaux-Dumez of France and Anglian Water Plc as well as a number of Argentinian companies. While privatisation did bring about improvements in drinking water, this was mainly in neighbourhoods already well catered for. The poorest areas remained badly serviced. However, it failed consistently to meet contract goals in terms of sewerage treatment, with sewerage effluents of over 5 million residents still being dumped into the River Plate. This was despite investment loans of more than US$911 million between 1993 and 1997 from the International Finance Corporation (IFC), the branch of the World Bank that invests in the private sector.

Serious charges have been levelled against the consortium for using debt payments to mask dividend returns to stockholders. In the face of public protest the consortium 'has shown utter contempt for public opinion and accountability'. There has been a general lack of transparency regarding its activities and performance, fuelling mistrust, while the consortium's monopolistic position affords it unilateral control of the data it releases into the pubic domain. The experience of Buenos Aires provided an early signal of some of the pitfalls associated with privatising water and sanitation systems. Given the overwhelming lack of faith by the majority of Latin Americans in the ability of market forces to resolve their problems, the international financial institutions have been forced to concede that privatisation in the continent has perhaps been oversold.

Source: Bosman (2005)

least profitable aspects of service delivery. This is how the 'hidden costs of privatisation' are passed on to government (Batley 1996: 749).

Graham and Marvin (2001) see private involvement in public services as promoting 'splintering urbanism', a process of selective privatisation, for example, of transport networks, communications or water supply, in which 'infrastructural networks are being "unbundled" in ways that help sustain the fragmentation of the social and material fabric of cities' (Graham and Marvin 2001: 33). Under such conditions, private firms cherry-pick the wealthier cities or regions or the better-off parts of a city, excluding poorer areas, fragmenting the urban infrastructure and exacerbating divisions between rich and poor (World Development Movement 2005). To the extent that a coherent system emerges it is at best characterised as 'fragmented coherence' (Van Horen 2004). Here the state, private sector and communities may all share responsibility for provision, with each compensating for the delivery shortfalls of the other, but with no clear mechanisms of accountability or coordination structures in place. The multiplicity of operators and the problems of integrating formal and informal service delivery networks increase transaction costs, while accountability black holes give rise to coordination difficulties for municipal governments.

The service delivery problems associated with NPM and the shape that they assumed under the rubric of urban management meant that the late 1990s saw the high water mark of privatisation. Sober reflection led to the recognition that while privatisation may enhance efficiency in some contexts it is clearly not always effective, and it is certainly far from equitable. The future privatisation of critical urban services is now uncertain, with increasing attention being paid once more to supporting struggling public utilities around the world. With 90 per cent of piped water in public hands, instead of bemoaning the fact that public utilities are inefficient, bureaucratic and unresponsive to the needs and demands of users, they should be supported so that they can deliver effective operations: 'We need to put an end to the traditional negative image of the public sector, and instead demonstrate an exciting, positive vision of "public-ness"' (Warwick and Cann 2007: 8). There is also a degree of pragmatism in this approach. However logical substituting the market for the state might have seemed, in practice it represented an enormous political challenge because governments had to implement the shift. In other words, government agents were supposed to collaborate in dismantling their own power and to cooperate in their own demise (Hirschmann 1993). In the end, the new urban management proved inadequate to the task of urban development and gave rise to as many problems as it was meant to rectify, starkly exposing some of the blind spots of 'mainstream development thinking' over the past few decades.

Urban transport and pollution

Although we have not classified transport as a critical urban service, a comprehensive and efficient transport system is also an essential ingredient for ensuring a mobile population and a healthy urban economy. Yet transport systems suffer 'fragmented coherence' as well – sometimes even incoherence. There are a number of reasons for this. Cities in low- and middle-income countries have experienced dramatic physical expansion with more people commuting longer distances than ever before. In Asia the city of Jakarta, Indonesia has experienced an explosion of peripheral settlements in the wake of commercial property development in the city centre and the demolition of more proximate informal settlements. In response, poor urban residents have moved to peripheral areas in search of cheaper land and housing and a lower cost of living: 'By the 1990s, over 70 per cent of Jakarta's population was gathering at the periphery, sometimes at a commuting distance of 120 kilometres' (Jellinek 2000: 272). While in much of Asia people depend on person-powered or motorised rickshaws, these distances underscore the fact that only small urban centres can function without motorised vehicular public transport.

As cities expand, public and private transportation networks have come under significant strain, with congestion, road accidents and vehicular emissions consequently on the rise. Economic liberalisation has led to increased wealth for some and this has been accompanied by a rise in the use of the private car (see Table 5.3). At the same time, structural adjustment policies (SAPs) saw cuts in expenditure on public transport,

Table 5.3 ***Trends in automobile usage in selected countries***

Country	*Motor vehicles per km of road*	
	1990	*2000*
Botswana	3	11
Ethiopia	2	3
Nigeria	21	14
South Africa	26	11
China	4	11
India	2	3
Philippines	4	11
Brazil	8	17
Chile	13	25

Source: UN-Habitat (2003b)

which led to a proliferation of unregulated forms of privately provided commuter transport, often unlicensed and informally organised. In Nigeria, for example, reduced public expenditure meant a decline in municipal bus services. At the same time public sector job cuts saw former government employees purchasing and running fleets of minibus taxis to make a living, taking advantage of the public transport gap (Oyeniyi 2007). Another way in which demand for affordable transport has been met is through a 'monumental expansion' of motorcycles in some Nigerian cities. For the urban poor experiencing declining living standards as Nigerian cities adjusted their economies, motorcycle taxis not only proved convenient, providing as they do a door-to-door service, but also were cheaper than minibus taxis (Bako 2007).

Nevertheless, throughout much of urban Africa, minibus taxis constitute the most common form of commuter travel along routes otherwise unserviced by public transport. It is the less well off urban residents who tend to live in peripheral settlements and work in distant city centres or commercial districts. While affordable and convenient, informally provided public transport is unregulated and often unsafe. Competition between private operators can be immense, resulting in tariff wars, overloading of vehicles and speeding to maintain a competitive edge. Poor urban dwellers using these services suffer the consequences. In Dar es Salaam, for example, 93 per cent of all fatal accidents in 1992 involved *daladala*, the name given to the privately operated Toyota minibuses, which constitute the city's main form of transport. A jovial response to the question 'How many people can a *daladala* hold?' is 'Two more!', while the names given to these taxis, such as 'Zig-Zag', speak volumes about the hazards involved as they compete for passenger loyalty by boasting of their ability to get to places quickly (Rizzo 2002: 144). This brings to life the statistics on road accidents provided by the WHO: globally road crashes are the second leading cause of death among people aged 5–29 years, and the third leading cause of death among people aged 30–44 years. Traffic injuries are therefore an important public health risk – a 'neglected epidemic' (Nantulya 2002). Furthermore, more than 85 per cent of all traffic deaths occur in low- and middle-income countries (Nantulya 2002). And the situation is getting worse. According to the WHO, 'without immediate action to improve road safety, it is estimated that road traffic deaths will increase by 80% in low- and middle-income countries by 2020' (WHO 2004). In cities, it is generally not those in cars who are at the greatest risk, but those on foot – invariably the least well off (see Plate 6). For example between 1977 and 1994, 64 per cent of

Plate 6 Crossing the street in Shanghai

Source: Tom Goodfellow

fatalities from traffic accidents in Nairobi were pedestrians (UN-Habitat 2006: 131).

Higher accident rates reflect a sharp increase in the number of vehicles on the road. Throughout much of Africa, Asia and Latin America the increase in the number of cars has far outstripped the building of new roads, with some notable exceptions. This is where there has been considerable investment in infrastructure. In India, which like Asia in general has seen some of the sharpest increases in deaths resulting from traffic accidents, the total number of motor vehicles more than doubled between 1990 and 2000 (Fazal 2006: 142). The explosion in the number of cars on the road 'is driven by powerful forces of inequality': part of 'a vicious circle in which the declining quality of public transport reinforces private auto use and vice versa' (M. Davis 2006: 131). The rapid increase of cars and motorways in cities without adequate public transport contributes to processes of social fragmentation that have knock-on effects that deeply entrench urban inequalities. Highways and motorisation 'have contributed to a coarsening, widening and stretching of the urban fabric' (Graham and Marvin 2001: 119).

In Managua, for example, there has been huge investment in new high-speed roads and roundabouts connecting parts of the city inhabited or used by elites. While these areas are now tightly interlinked, those areas deemed insignificant or dangerous are bypassed and systematically neglected (see Box 5.4). As a result, the city has become safer for the rich – both because of the quality of the roads and because they have increasingly little contact with the poorer and more threatening parts of the city – while the urban space available to the poor has become increasingly circumscribed as 'large swathes of the metropolis' are carved out 'for the sole use of the urban elites' (Rodgers 2004: 123). These processes are by no means exclusive to low- and middle-income countries and have been widely noted in relation to North American cities. However, in many cities of Africa, Asia and Latin America, persistent poverty and low public investment put safe and affordable mobility more firmly out of reach for a larger number of urban residents.

Transport specialists are now largely agreed that for most large cities a bus-based mass transit system is the most cost-effective option. The city

Box 5.4

'Disembedding' Managua

Since the late 1990s, Managua, the capital city of Nicaragua, has undergone an extensive process of urban development that has underpinned profound socio-spatial segregation. Through a mixture of both public and private initiative, the urban fabric of what was previously a relatively egalitarian metropolis has been completely transformed, with an elite-oriented 'Nueva Managua' emerging in the form of luxurious housing developments, opulent malls, and the proliferation of expensive restaurants, bars and nightclubs around the city. These have been linked together into a veritable 'fortified network' through the selective construction of well-maintained, high-speed roads and roundabouts across the metropolis that allow for members of the urban elite to move safely and speedily between the different spaces of their lives. The poor are very visibly excluded not only from the new gated communities, malls and bars by an ever-expanding number of private security guards, high walls, and modern surveillance technology, but also from the new connecting road network. This is cruised at breakneck speeds by expensive 4×4 cars, and have no traffic lights but only roundabouts, meaning that those in cars do not have to stop – and risk being carjacked – but those on foot risk their lives every time they try to cross a road. The general picture, in other words, is one whereby a whole 'layer' of Managua's urban fabric has been purposefully 'disembedded' from the general patchwork quilt of the metropolis for the exclusive use of the elite, thereby profoundly altering the cityscape and the relations between social groups within it.

Source: Rodgers (2004)

of Curitiba, Brazil provides a paradigmatic example. The city developed a bus-based public transport system in the 1970s. It is a centralised and integrated system rather than a series of competing routes, although private bus companies operate them. Dedicated bus-ways were established from the outset so that road capacity was given over to buses rather than cars, and speed is facilitated by passengers being able to pay for their trips before boarding the bus, which they do so very rapidly. Buses are therefore able to move people from one area to another as swiftly as the New York subway but at a fraction of the cost (Tannerfeldt and Ljung 2006: 101–102). Another success story is Bogotá, the capital of Colombia, which has introduced a bus-based rapid transit (BRT) system across the city. Known as the TransMilenio, it was initiated in 1999 by way of a public–private partnership that offers frequent and affordable transport across the capital. The backbone of the system is a high quality bus network operating on over 80 kilometres of dedicated bus lanes catering for 850 articulated buses. The BRT system traverses rich and poor parts of the city alike, with free feeder buses ferrying passengers to the 114 TransMilenio stations, also served by 250 kilometres of cycle paths.

In a city of over 8 million people, congestion has improved dramatically, as has the quality of life of the majority of Bogotá's inhabitants. An average trip now takes one-third of the time it used to and this combined with service frequency means that the buses are 95 per cent full as opposed to less than half full in the past. Accidents have decreased by 80 per cent and air pollution has dropped almost 40 per cent. TransMilenio's success has been facilitated by a good system of metropolitan taxation and investment, as well as a positive change in attitude among the city's leaders and inhabitants. The determination and political will of two former mayors of Bogotá – Enrique Peñaloza and Antanas Mockus – resulted in more equitable city planning and a drive to educate citizens through a popular and successful public campaign. Unfortunately, initiatives such as the TransMilenio in Bogotá are rare, and even when public transport is improved it does not stop the wealthy from using their private cars. In low- and middle-income cities per capita vehicle ownership is generally lower than in North America and Europe for example. However, the state of the vehicles means they often do more damage, as governments in many poorer countries allow for much higher levels of lead and other pollutants in fuel than is permitted in the advanced economies.

Vehicular emissions, alongside emissions from factories and the burning of low-grade domestic fuels, generate atmospheric pollution that has both local and global effects. Cities in advanced economies account for 65 per cent of the world's resource use and waste production. In the past, most urban air pollution in low- and middle-income countries came from industrial activity, power stations or domestic burning of coal and heavy oil. Nowadays cities in these countries have joined the advanced economies in seeing road transport as a significant source of pollutants as well. The predominant sources of air pollution vary from city to city even within the same country. In India, for example, vehicle emissions contribute most to air pollution in Delhi, while in Kolkata domestic coal-burning stoves are the largest source (Maitra and Krishan 2000: 184). While it is difficult to isolate the effect of one particular pollutant from others, and each city has its own particular proportions forming a unique toxic cocktail, it is estimated that half a million deaths worldwide can be attributed to particulate matter and sulphur dioxide alone. In India it is thought that air pollution causes 40,000 premature deaths each year, while an estimate for China suggests that smoke and particles from burning coal cause more than 50,000 premature deaths and 400,000 new cases of bronchitis in eleven of its largest cities (Hardoy *et al.* 2001: 101). Lead is another major concern, with one study from Mexico City in the late 1980s finding that over a quarter of newborn infants had lead levels in their blood that were high enough to impair neurological and motor-physical development (Hardoy *et al.* 2001: 103). Clearly well-managed and regulated transport systems must play a role in reducing problems of atmospheric pollution.

The WHO estimates that 1.5 billion urban dwellers worldwide are exposed to levels of ambient air pollution above recommended maximum levels (Hardoy *et al.* 2001: 90). For example, Beijing has over four times as much particulate matter as London, and almost four times as much sulphur dioxide; while Guangzhou has an astonishing seventeen times the amount of sulphur dioxide as London (UN-Habitat 2006: 221–223). The annual mean concentrations of ambient particulate matter in Indian cities is more than six times the urban mean of the United States, and there are also serious problems with carbon monoxide, nitrogen oxides and hydrocarbons emitted by vehicles (Hardoy *et al.* 2001: 99). Indeed, breathing Mumbai's air is said to be the equivalent of smoking twenty cigarettes a day (Abhat *et al.* 2005), while many of the smaller but rapidly growing cities in India, such as Saharanpur, 'are on the threshold of an environmental crisis due to growing air pollution' (Fazal 2006: 150). Sufficient levels of these pollutants can precipitate a range of major health problems as well as substantial rates of mortality.

It has to be wondered just how choked our cities must become before significant and widespread action is taken.

Atmospheric pollution acts as something of a social leveller. Both rich and poor ultimately breathe in the same air. However, as Marianne Kjellén of the Stockholm Environment Institute has observed: 'The environmental burdens of poverty are suffered by the poor and dealt with by the poor. The environmental burdens of affluence are suffered by the public and dealt with by the government' (cited in Tannerfeldt and Ljung 2006: 62). Moreover, it is the activities of the wealthy that invariably contribute to deteriorating air quality and the urban poor who have little control over the impact. As such, equity remains a salient issue when considering the risk posed by urban transport and atmospheric pollution. Both clearly present urban planners with complex challenges – especially considering that these issues stand at the intersection between environmental concerns on the one hand and the imperatives of poverty reduction and development on the other. Given the many links between the experience of urban poverty and the quality of the urban environment, an integrated approach that addresses both problems seems obvious. This has begun to emerge under the rubric of sustainable development.

Sustainable cities and the global environment

Manuel Castells has remarked that 'Our blue planet is fast becoming a predominantly urban world and in this context, *sustainable development* is the code word for the most important social debate of our time' (Castells 2002: ix–x). Cities are at the heart of debates on sustainable development, although confusion still exists as to what it means. As originally formulated by the World Commission on Environment and Development (WCED) (also known as the Brundtland Commission) sustainable development refers to: 'Development that meets the needs of the present, without compromising the ability of future generations to meet their own needs' (WCED 1988). Yet nowadays there are as many definitions of sustainable development as there are designations of development. Disagreements arise over what it is that is to be sustained. At least two differences in interpretation can be identified, depending on where emphasis on sustainability is placed: a sustained meeting of human needs, or keeping natural resources and ecosystems intact (Mitlin and Satterthwaite 1996; Satterthwaite 1999). The first, the sustained meeting of human needs, relates in part to what has become known as the 'brown agenda'. This refers to the urban environment as it affects urban dwellers:

the provision of safe water, sanitation and drainage, and solid and hazardous waste management. The brown agenda emerged as a matter of concern because it was perceived that the environmental movement neglected the local impact of urban environmental problems, especially in low- and middle-income countries. It can also include atmospheric pollution, although this aspect of the urban environment is sometimes dubbed the 'grey agenda'.

The second emphasis on environmental sustainability – sometimes referred to as the 'green agenda' – focuses on conservation, biodiversity and the protection of natural resources. The management of the urban environment clearly has important impacts on broader environmental concerns and the failure of many cities to address this has led environmental activists to cast cities in the role of ecological culprits. However, there are other ways of looking at the relationship between cities and environmental sustainability. While the concentration of people and activities in urban centres does lead to high levels of pollution, that very same concentration offers opportunities for environmental management strategies that reduce environmental degradation, for example through the recycling and reuse of resources. As UN-Habitat has pointed out:

> [Cities] represent a much greater potential for limiting the use of motor vehicles, including greatly reducing the fossil fuels they need and the air pollution and high levels of resource consumption that their use implies. This might sound contradictory, since most of the world's largest cities have serious problems with congestion and motor-vehicle generated air pollution. But cities ensure that many more trips can be made through walking or bicycling. They also make possible a much greater use of public transport and make economically feasible a high-quality service.
>
> (UNCHS 1996: 419)

Further, urban areas contain around half the world's population and yet cover only a small fraction of the total land area of the planet. As such they have the potential to limit further potential environmental damage and resource depletion (Marcotullio and McGranahan 2007).

Nevertheless, tensions exist between brown agenda advocates and those concerned with wider environmental concerns (or 'green issues'). For a start, solutions to brown agenda dilemmas can be problematic from a green perspective. Improved water supply and sanitation in cities can have knock-on effects that are damaging to the wider regional or global environment. For instance, if a city manages to create a widespread functional waterborne sewage system, thereby massively reducing the exposure of its residents to faecal matter and biological pathogens, its demand for water escalates dramatically, increasing the drain on water resources in the region of which it is a part. In other words, solving a brown problem is not necessarily a step in the right direction towards solving green movement concerns. It is often noted that cities have large 'ecological footprints'. This term is used to reflect the combined use of renewable resources by urban dwellers beyond the physical boundaries and carrying capacity of a city (Rees 1992).

It is certainly true that cities can have devastating effects on their surrounding regions. For example, meeting the needs of urban dwellers for firewood can speed up deforestation, while cities can use up the lion's share of limited water sources in a region. Similarly, poor urban sanitation can contaminate rivers with the consequences being felt downstream and hundreds of miles beyond the urban edge. The damage done is not only to people's health but also to their livelihoods. Sea and river fish yields can decline as a result of water pollution from activities in nearby cities. Air pollution has a huge regional impact, with acid rain often falling hundreds of kilometres away from the source of pollution, devastating agriculture by ruining soil and vegetation. The expansion of cities on to previously rural land can push agriculture into less suitable areas, while the commercialisation of agriculture responding to need stemming from cities renders many small farms redundant. The natural demands of cities increasingly shape the environment of their surrounding regions; indeed, while many cities first developed as market centres to serve the farms and farming households around them, urban hinterlands now increasingly work to serve the city.

There are certainly grounds for taking a dim view of cities from an environmental perspective. However, as Tannerfeldt and Ljung (2006: 66) have observed, it is increasing affluence and the production and consumption associated with economic growth rather than the spatial configuration of cities that leaves the footprint, even if economic development cannot take place without cities. Furthermore, many brown agenda problems are perpetuated neither by urban spatial dynamics nor rapid economic growth, but rather by chronic underinvestment and neglect in addressing persistent poverty in cities. In many ways brown agenda issues reflect the consequences of underdevelopment, while green issues reflect the opposite: the results of rising prosperity and high-level mass consumption. Nevertheless, rigidly separating the two agendas, or dealing with the issues of one without paying adequate attention to the other, can be counterproductive. It is too simplistic to interpret brown issues in relation to development and green issues to the environment. To do so risks undermining the positive achievements of linking the two. It is for this reason that the idea of sustainable development has become so significant, in many respects constituting an attempt to reconcile and harmonise the two agendas.

Tackling poverty and sustainability in cities while simultaneously locating urban issues within a broader environmental perspective has to be at the heart of urban environmental management both on behalf of people living and working in cities and the global population at large. The combined impacts of expanding industrial production, poor transport networks and vehicular emissions in low- and middle-income countries are contributing an increasing proportion of global greenhouse gas emissions. The consequences are acute at the local level, but also have implications for global efforts to achieve environmental sustainability. Consequently, effective urban management has broad implications for the global community.

Nonetheless, it is important to recognise that the relationship between cities and the global environment works in two directions. Cities are vulnerable due to a failure to address climate change at the national and international levels. This is not to imply that rural areas should not receive attention. Indeed their vulnerability should be seen and responded to alongside urban vulnerability. Yet a particular argument for climate change action in relation to urban centres in low- and middle-income countries has to be made for three critical reasons (Reid and Satterthwaite 2007: 1):

- They already have three-quarters of the world's urban population; China's urban population is as large as Europe's; India and Africa both have larger urban populations than North America.
- They will house most of the growth in the world's population over the next ten to twenty years, and have a major impact on future greenhouse gas emissions.
- They have a large and growing proportion of the world's population most at risk from storms, floods and other climate change-related impacts.

Within these populations those most at risk are in low-income households because they live in poor quality dwellings located in areas most vulnerable to threat. They are also those least able to escape disaster, cope with hazards or recover from these and other impacts of climate change.

Urban centres more generally are important to protect as they are host to between 60 and 95 per cent of economic activities in low- and middle-income countries (Huq and Satterthwaite 2008: 1).

The devastation being brought to urban populations as a result of extreme weather events is increasingly evident, with both young and elderly people being susceptible to risk from heatwaves and people being killed or harmed by storms, floods, drought and landslides. Even when these are not directly the result of climate change, 'it is proof of the vulnerability of urban populations to floods and storms whose frequency and intensity climate change is likely to increase' (Satterthwaite *et al.* 2007: vii). Particularly vulnerable are urban centres located in low-elevation coastal zones that are less than 10 metres above sea level. These areas, which comprise less than 2 per cent of the world's land area, contain 10 per cent of its total population (over 600 million people) and 13 per cent of its urban population (around 360 million people): 'Almost two-thirds of the world's cities with more than five million inhabitants fall in this zone, at least partly', and low- and middle-income countries have nearly twice the proportion of their urban populations in this zone (Satterthwaite *et al.* 2007: 22). Examples of cities already identified as vulnerable are Dhaka, Bangladesh (see Box 5.5); Mumbai and Shanghai with parts of both being only 1–5 metres above sea level; Alexandria, Port Harcourt and Banjul in Africa; New Orleans in the United States and the coastal towns of the Caribbean states, where between 20 and 50 per cent of the population reside close to sea level (Satterthwaite *et al.* 2007: 24). The most dramatic case is China, which has the largest number of urban and rural dwellers in the low-elevation coastal zone and where the coastal provinces experienced a net in-migration of around 17 million people between 1995 and 2000 due to increasing economic opportunities (McGranahan *et al.* 2007).

Absolutely clear when assessing the actual and potential harm to urban populations of climate change is that solutions cannot be left to local government, or even large metropolitan authorities. Urban institutions can take actions to provide services, conserve water, recycle waste and reduce greenhouse gas emissions. However, they cannot be held responsible for reducing climate change risk beyond their jurisdictions. What urban local governments can do, given that cities cannot move, is to work on their adaptive capacity, being 'the *potential* of a system or population to modify its features or behaviour to cope better with existing and anticipated stresses' (Satterthwaite *et al.* 2007: 50). This involves both planned interventions and systems, both reactive and anticipatory, such as

Box 5.5

The impact of climate change on Dhaka, Bangladesh

Dhaka has over 10 million inhabitants and has been central to Bangladesh's economic success in recent years. The city is already very vulnerable to flooding, especially during the monsoon season, as shown by five major floods since 1980. The 1988, 1998 and 2004 floods were particularly severe and brought large economic losses. When major floods occur, they hit around half the city's area. They are mainly caused by spillover from surrounding rivers and rainfall that generates runoff beyond the capacity of the drains. The provision for infrastructure and services has not kept pace with the city's population growth. The 40 per cent of the population living in slums and squatter settlements are most severely affected by floods, waterlogging and other associated problems. The unequal development and management of utilities and bad management of water and waste water are caused by both non-compliance with national policies, rules and regulations and insufficient resources.

Climate change will affect Dhaka in two main ways: through floods and drainage congestion and through heat stress. Melting glaciers and snow in the Himalayas and increasing rainfall will lead to more frequent flooding (waterlogging, drainage congestion from river floods and excessive rainfall during the monsoon already cause very serious damage). Furthermore, Dhaka may also face 'heat island' problems, because temperatures in the city are a few degrees higher than in surrounding areas. Infrastructure, industry, commerce and utility services are key sectors at risk from flooding. The impaired performance of these sectors during and after flooding increases the vulnerability of city dwellers, with economic and social class dictating the severity of that vulnerability. The loss of life and livelihoods, and impacts on human health, are particularly severe for lower-income groups.

Source: Alam and Rabbani (2007); Huq and Satterthwaite (2008)

rapid restoration of infrastructure or adapting land-use planning and regulatory frameworks to reduce the vulnerability of urban dwellers. It also means being responsive to the spontaneous adaptations made by individuals and groups within cities. For example, a 'Climate Future for Durban Programme' was initiated with the South African Council for Scientific and Industrial Research (CSIR). In 2006 the report produced a 'Headline Climate Change Adaptation Strategy' for the city, which includes predicted changes and likely effects of long-term climate change, the likely effects on infrastructure, water and sanitation, transport, human health, food security, agriculture, business and tourism, biodiversity and the coastal zone. Seeing the city within its broader regional context,

Durban's Metropolitan government has also identified the implications of climate change for each municipal department and has sought to build on the existing capacity in relation to disaster-risk reduction to enhance preparedness and adaptability to climate change (Satterthwaite *et al.* 2007: 56–58).

While local knowledge and capacity is critical in facilitating mitigation and adaptation at the micro- and meso-levels, responsibility cannot rest here. For a start, the costs are potentially infinite. Oxfam estimates that a minimum of US$50–80 billion annually is necessary for low- and middle-income countries to adapt to climate change (Oxfam 2007). Competence and capacity demands are extremely high, while the waste resulting from a failure to embark on joined-up approaches that link cities and their hinterlands would be errant. Hence support from higher levels of government is essential. However, this is also a matter of global concern. Low- and middle-income countries and their urban centres do not have the power and resources to address the profound global inequities that caused climate change in the first place, even if they are home to those most vulnerable to its impact. While the world is waking up to climate change and responding, however slowly, incrementally and iteratively, the plight of cities remains fairly low down the agenda of aid donors and international agencies.

Conclusion

In thinking about and trying to address the challenge of urban poverty and vulnerability, it is important to look beyond the most visible manifestations – the poor quality of dwellings and the legions of informal workers, producers and traders that are so often in evidence in the streets of cities. While livelihoods and housing are critical, the integrity of the built environment of a city is of equal concern. Deficient infrastructure raises the costs of acquiring basic necessities such as water, and creates a more dangerous disease environment, which can have far-reaching effects on the ability of the poor to improve their lot. Poorly designed and managed transport infrastructure slows down the flow of people and goods in a city, while also contributing to further health risks such as traffic accidents and air pollution. Air pollution contributes to the global threat of climate change – a threat to which the poor are most vulnerable. From the household level to the global level, the effective management of urban environments is essential.

There is, however, another class of environmental risk that is not often considered as such. The threat of violence – in the form of crime, social conflict or war – is a very real environmental hazard for millions of urban dwellers, both rich and poor. In Chapter 6, we explore this issue with particular reference to the challenges of crime, war and terrorism, and – conversely – highlight the potential that well-governed cities have to mitigate these pernicious forms of vulnerability.

Summary

- **Water supply, sanitation and solid waste management are considered *the* critical urban services. Hundreds of millions of people worldwide lack access to adequate water supplies and even more lack adequate sanitation.**
- **The management of solid waste is an often neglected aspect of urban services. It is a stark indicator of inequality, given that most waste is created by the wealthy but reused and recycled by the poor.**
- **Delivering services in highly unequal cities poses questions about how to upgrade services in low-income areas and prevent the creation of permanent underserviced ghettoes, where poor services impact on health, education and social mobility.**
- **The neo-liberal turn in the 1980s saw the introduction of 'new public management', whereby governments were conceived less as deliverers of services than as managers of private contractors, in the interests of greater efficiency and choice.**
- **In this context many urban services were conceptualised as private, economic goods rather than public goods. Ultimately this approach produced greater inequality in service delivery in most cases and further fragmentation of urban infrastructure.**
- **Urban transport increasingly has been provided by private and informal actors as well. Combined with sharp increases in numbers of vehicles, this has led to rises in both deaths on the road and urban atmospheric pollution.**
- **Questions of urban services and the immediate urban environment (the 'brown agenda') are linked to wider and longer-term environmental concerns (the 'green agenda') through the idea of 'sustainable development'.**
- **Cities are also closely linked to climate change, but the relationship is two-directional: while urban consumption is a significant contributor to global warming, poor urban dwellers are also among the most vulnerable to the effects of climate change.**

Discussion questions

1. Why has it proven so difficult to rectify deep inequalities in the delivery of critical urban services?
2. Discuss some of the potential benefits and problems associated with privatising urban water supplies.
3. Are all urban services best conceptualised as 'public goods'?
4. Discuss the tensions and complementarities between the 'brown' and 'green' agendas.
5. To what extent can local and metropolitan government alone limit the environmental damage caused by urban production and consumption?

Further reading

Batley, Richard and George Larbi (2004) *The Changing Role of Government: The Reform of Public Services in Developing Countries*, Basingstoke: Palgrave Macmillan.

Davis, Mike (2006) *Planet of Slums*, London: Verso.

Graham, Stephen and Simon Marvin (2001) *Splintering Urbanism: Networked Infrastructures, Technological Mobilities and the Urban Condition*, London: Routledge.

Hardoy, Jorge, Diana Mitlin and David Satterthwaite (2001) *Environmental Problems in an Urbanizing World: Finding Solutions in Cities in Africa, Asia and Latin America*, London: Earthscan.

Marcotullio, Peter and Gordon McGranahan (eds) *Scaling Urban Environmental Challenges: From Local to Global and Back*, London: Earthscan, pp. 1–17.

Satterthwaite, David (2006) 'Climate change and cities', *IIED Sustainable Development Opinion*, London: International Institute for Environment and Development.

Tannerfeldt, Göran and Pers Ljung (2006) *More Urban Less Poor: An Introduction to Urban Development and Management*, London: Earthscan.

UN-Habitat (2003) *Water and Sanitation in the World's Cities*, London: UN-Habitat and Earthscan.

Useful websites

British Council site on the relationship between cities and climate change: www.britishcouncil.org/zerocarboncity-cities.htm

International Institute for Environment and Development (IIED): www.iied.org

Water Supply and Sanitation Collaborative Council (WSSCC): www.wsscc.org

World Water Council: www.worldwatercouncil.org/index.php?id=23

6 Human security in cities: crime, violence, war and terrorism

Introduction

Crime, endemic violence, war and terrorism are among the most dramatic and disturbing manifestations of vulnerability in cities. And cities are uniquely prone to these threats given their economic, political and cultural significance. Historically, many cities were designed and planned with this vulnerability in mind. From the ziggurats of Mesopotamian city-states, to medieval fortress towns in Europe, to fortified colonial outposts, the impetus for city-building has often been inspired by a dual desire to concentrate and project power, as well as to defend it against perceived external threats. But as spaces where political and economic power is concentrated, where diverse actors converge, and inequality is highly visible, threats just as often emerge from within as without. This too has not been lost on city-builders: the grand boulevards of Paris were not designed for promenading, but rather military parading, to emphasise the power of the state and ensure that military order could be maintained in the streets if insurrection should materialise.

Despite the intrinsic links between cities, city-building and conflict, the relationship between these factors has received only scant attention in the context of development. Yet crime, violence, war and terrorism present very real and very serious obstacles to development in cities and beyond. We begin this chapter by outlining the concept of 'human security', which emphasises the link between the security of ordinary people everywhere and the broader concern with 'global security'. Human security comes under threat in many ways. The most pervasive menaces are crime and endemic violence, which are particularly acute when public authorities fail to maintain law and order in city streets and violence becomes an organised pursuit of criminal, insurgent or terrorist networks. Tackling crime and endemic violence in cities is especially challenging in contexts of rapid urbanisation and political transitions. Indeed, transitions from war to peace are frequently linked to a spike in urban crime and violence, with some authors arguing that we are witnessing a historical shift in the geography and manifestations of socio-political conflict. This is especially apparent in the case of 'new wars', which involve 'asymmetrical' tactics such as terrorist attacks designed to maximise civilian deaths and inculcate fear. As cities emerge from armed conflict – whether in the form of gang warfare, communal conflict, terrorist attacks or outright war – urban reconstruction poses not only complex challenges, but also unique opportunities. Indeed, cities can be important sites for cultivating peace and building a concrete image of a society's future aspirations.

Human security and development

There has been a growing tendency in recent years to link global security concerns to international development. War, terrorism and organised crime are depicted as both evidence of development failure and obstacles to development. In an increasingly globalised world – it is argued – the effects of poverty and insecurity are no longer strictly confined by national boundaries (see Box 6.1). As Peter Liotta (2007) points out:

> Anarchy, governmental collapse, ethnic rivalry, cultural grievances, religious-ideological extremism, environmental degradation, natural resource depletion, competition for economic resources, drug trafficking, alliances between narco-traffickers and terrorists, the proliferation of 'inhumane weapons' and the spread of infectious diseases threaten everyone. It is not possible to be isolated from their effects.
>
> (Liotta 2007: 12–13)

This tendency has raised some concern among development scholars and professionals who fear that conventional development goals, such as reducing poverty and inequality, will be eclipsed by apparently more pressing national security concerns (Beall *et al.* 2006). There is, however, no question that security and development are intrinsically linked. Indeed, *insecurity* – or vulnerability – is a defining characteristic of poverty.

Box 6.1

Security, development and US foreign policy

Following the events of 11 September 2001, the United States revised its national security strategy, famously asserting a right to strike 'pre-emptively' in defence against perceived national security threats. However, the strategy also devotes a chapter to development, which is also presented as a 'pre-emptive' measure:

> Helping the world's poor is a strategic priority and a moral imperative. Economic development, responsible governance, and individual liberty are intimately connected . . . The United States must promote development programs that achieve measurable results – rewarding reforms, encouraging transparency, and improving people's lives (31).
>
> Development reinforces diplomacy and defense, reducing long-term threats to our national security by helping to build stable, prosperous, and peaceful societies. Improving the way we use foreign assistance will make it more effective in strengthening responsible governments, responding to suffering, and improving people's lives (33).

While this is a welcome recognition of the importance and imperative of development, it also raises concerns that military priorities and approaches to 'development' will overshadow non-militarised interventions.

Source: US National Security Strategy (March 2006)

Whether or not the push to link security and development proves beneficial largely depends upon how 'security' is conceptualised and pursued.

In the Cold War era, security was synonymous with sovereign integrity. In other words, security was conceptualised fairly narrowly 'as security of territory from external aggression, or as protection of national interests in foreign policy or as global security from the threat of a nuclear holocaust' (UNDP 1994: 22). In the quest to maintain sovereign integrity, 'development assistance' was frequently delivered in the form of military aid to governments: 'forgotten were the legitimate concerns of ordinary people who sought security in their daily lives' (UNDP 1994: 22). At best, the concerns of ordinary citizens were overlooked in the pursuit of a stable world order; at worst, the primacy of international security concerns contributed to greater insecurity in the daily lives of ordinary citizens. In the post-Cold-War world, where intra-state conflicts are more common that inter-state warfare, and no nation seems immune to the threat of terrorism, the concerns of ordinary citizens are attracting increasing attention. Indeed, the changing global political landscape following the collapse of the Soviet Union inspired the United Nations Development Programme to promote a new concept of security – *human security* – that places people, not nation-states, at the centre of the agenda.

Human security is a holistic concept that embraces a wide range of concerns, including economic security, food security, health security, environmental security, personal security, community security and political security (UNDP 1994: 24–25). In other words, the concept of human security is fundamentally motivated by 'a concern with human life and dignity' (UNDP 1994: 22). If this list of security objectives seems impossibly long, it is worth noting that 'International peace and security is ultimately constructed on the foundation of people who are secure' (Human Security – Cities 2007: 10–11). The steady flows of migrants moving from poor countries to rich countries in search of better opportunities, the acceleration of climate change, and the recent rise in terrorist attacks from New York City to Bali all attest to the increasingly globalised nature of threats to human security.

By placing the security concerns of people over those of nation-states, the concept of human security also highlights the intrinsic relationship between human security and development. Indeed, 'human safety and security . . . is increasingly being acknowledged internationally as a public good, as well as a basic human right' (UN-Habitat 2007: 45). We have seen the consequences of, and challenges associated with tackling, several dimensions of human insecurity in cities where states and markets fail to serve the needs of the poor (such as sustainable livelihoods, adequate housing and environmental security). Here we turn our attention to the most pernicious forms of insecurity experienced by those living in cities: crime and violence.

Most definitions of violence begin with the application of physical force, although broader classifications include the presence of psychological hurt, material deprivation and symbolic disadvantage. However, all definitions recognise that 'violence involves the exercise of power' (Moser and Rodgers 2005: 4). Broadly speaking, analyses of violence break the phenomenon down into economic, social and political forms. Economic violence is usually associated with crime, both organised and sporadic, and ranges from drug trafficking, car hijacking and kidnapping, through to mugging and armed robbery, small-arms dealing and petty theft. Social violence is more difficult to isolate. At one extreme it includes domestic violence in the form of physical or psychological abuse, and at the other, interpersonal violence in public places such as drunken brawls, arguments getting out of hand and road rage. Political violence also takes many forms, from riots to rebellions to terrorist attacks, but essentially involves the use of violence to challenge existing power relations. However, these various forms of violence often overlap, rendering a simple typology of the *motivations* for violence problematic.

For example, in 2008 a contested election in Kenya resulted in clashes between ethnic groups supporting opposing candidates. When violent political clashes fall along ethnic lines, is this political violence or social violence? When people take to the streets and riot in protest against rising prices for water due to the privatisation of provision, is this economic violence or political violence? Even when urban riots and insurrections have local causes rooted in economic concerns, they can have wider political implications (Hobsbawm 2005). Furthermore, the predominant manifestation of violence in a society may change over time while the underlying causes remain fairly constant. Anthropologist Dennis Rodgers (2007a) has argued that rising urban violence in Central America is not a qualitatively new form of conflict but a continuation and relocation of earlier – and ostensibly more political – forms of violence motivated by discontent surrounding gross inequalities in wealth. Revolutionary struggles have given way to endemic urban violence: 'although past and present forms of brutality might initially seem very different, present-day urban violence arguably represents a continuation of past struggles in a new spatial context' (Rodgers 2007a: 3). Establishing the exact nature and origins of the shift in urban violence over the past few decades illustrates once again the problems encountered when attempting to distinguish between 'political' or 'social' or 'economic' forms of violence.

Given the difficulty of classifying violence in terms of motivations, it is perhaps more useful to consider the *manifestations* of violence – particularly when considering the effects of violence on human security. Urban violence manifests in many forms, from isolated crimes to outright warfare (see Box 6.2). At one end of the spectrum lies 'anomic' violence – acts committed by individuals such as assault, murder, gender-based violence and theft on an ad-hoc basis. While such acts contribute to insecurity, generally speaking anomic violence does not constitute a serious threat to human security. It is when violence becomes an organised pursuit that it presents the most severe threat to individual, community and national security (Human Security – Cities 2007: 23).

Box 6.2

Typology of urban crime and violence

Anomic crime

Main actors: Individual criminals, state security/police forces.

Organisational features: Ad-hoc acts of violent crime and delinquency, usually economically motivated.

Impacts/outcomes: Sporadic murder, assault, gender-based violence, robbery/theft.

Organised crime

Main actors: Drug cartels, human trafficking networks, arms smugglers, state security forces (intelligence) and police officers.

Organisational features: Command structure, often transnational, limited territorial control, mainly economically motivated.

Impacts/outcomes: Targeted killings, kidnapping, extortion, systematic sexual abuse, human trafficking and enslavement, small arms proliferation.

Endemic community violence

Main actors: Urban gangs, vigilante groups/community defence organisations, ethnic militias, state security forces and police officers.

Organisational features: Widespread/routine violent crime in the context of failed public security, limited command structure and territorial control, primarily economically motivated.

Impacts/outcomes: High rates of gang/police/civilian casualties, unlawful killings, recruitment of 'urban child soldiers', social cleansing, gender-based violence, inter-gang warfare and police shoot-outs, kidnapping, trafficking, robbery/theft.

Open armed conflict

Main actors: Rebel groups, paramilitaries, state military forces.

Organisational features: Struggle for territory in interstate or civil war context occurring in cities, usually large scale, political/ideological/identity motivated.

Impacts/outcomes: Significant civilian casualties, mass population displacement, war crimes and crimes against humanity, genocide, terrorism, humanitarian crises, gender-based violence, recruitment of child soldiers.

Source: Human Security – Cities (2007: 16)

Organised violence includes everything from vigilante groups to criminal gangs to terrorist networks to formal armies or militias. It is the activities of these groups that present the most grievous threat to human security in cities.

In our rapidly urbanising world, incidents of organised violence are increasingly an urban phenomenon – the 'peasant wars of the twentieth century' (Wolf 1969) have given way to 'urban wars of the twenty-first century' (Beall 2006). Crime rates are rising; social conflicts with historical roots in rural areas are being transferred to urban centres; cities are increasingly the primary battle field for 'new wars'; and terrorists use cities both as bases and targets for their activities. Even in countries where conflict remains primarily focused in rural areas, there are important spillover effects into cities, which become spaces of refuge for displaced populations (Beall 2007).

Given these trends, the pursuit of human security in cities must form an integral part of the development agenda. While urban authorities – such as politicians, planners and police forces – have some control over these factors, they cannot combat them alone. Gross inequality, international political grievances and transnational networks of criminals or terrorists are often symptoms of social, political, economic and historical factors that originate beyond city boundaries. To truly achieve and maintain human security in cities therefore requires a significant shift in governance strategies that involves cooperation from the very local to global levels, engaging communities, city authorities, national governments and international agencies.

Urban crime and endemic violence

Crime-related violence is the most pervasive form experienced by urban residents. In the year 2000, for example, there were approximately 520,000 deaths worldwide due to homicide, and approximately 310,000 war-related deaths (UN-Habitat 2007: 68). When assault, theft, human trafficking and interpersonal violence are factored in, millions of people are directly affected by some form of crime or violent incident every year. Furthermore, crime has been rising: between 1980 and 2000, global crime incidence increased approximately 30 per cent. While not all forms of crime are necessarily violent (e.g. theft) and not all forms of violence are criminal (e.g. domestic abuse, which is legal in many places), all forms of crime and violence contribute to human insecurity and undermine the basic human rights to safety and security of person and property.

The consequences of urban crime and endemic violence, however, are multidimensional and multispatial. While the negative effects for the individuals affected are clear, crime and violence have broader implications. At the local level crime and violence can undermine the cohesion and solidarity of community. At city level, endemic insecurity can generate a culture of fear leading to segregation, social fragmentation and the creation of fortified spaces. At national level, high rates of crime (or a perception of pervasive crime) can impact negatively on economic development. And at the international scale, lawless cities often provide ideal environments for transnational criminal organisations dealing in drugs, arms, and human beings.

In the Latin American context, which (as a region) has the highest homicide rates in the world, 'widespread fear and insecurity [are] a fact of daily life' (McIlwaine and Moser 2007: 120) and contribute to 'social erosion' (McIlwaine and Moser 2004) by undermining community and intergroup relations and leading to the atomisation of urban social life. The following quote from Don Sergio, a resident of a low-income neighbourhood in Managua, Nicaragua illustrates the point:

> Nobody does anything for anybody anymore, nobody cares if their neighbour is robbed. Nobody does anything for the common good. There's a lack of trust, you don't know whether somebody will return you your favours, or whether he won't steal your belongings when your back is turned. It's the law of the jungle here.
>
> (Rodgers 2006: 271)

The profound sense of insecurity expressed by Don Sergio affects millions of residents in cities across Latin America and Africa and Asia as well. Cities such as Rio de Janeiro, Lagos, Nairobi and Johannesburg are internationally renowned for crime and endemic violence – and their reputations are often justified.

While such high levels of violence are very serious, the anxiety generated by the perception of crime and endemic violence often outstrips the actual level of danger. And fear has social and developmental consequences as well: 'Fear, just like crime, can be portrayed as having damaged the fabric of cities, to have adversely affected the quality of urban life' (Bannister and Fyfe 2001: 808). In Mexico City, widespread acts of violence not only impact on the city directly, but also haunt the minds of its residents: insecurity has become 'a "phantom" that wanders the metropolis'

Table 6.1 ***Percentage of population feeling unsafe on the street after dark***

City (country)	*%*
Rome (Italy)	44
London (England)	42
Sydney (Australia)	27
Paris (France)	22
New York (USA)	22
Amsterdam (Netherlands)	22
Hong Kong (SAR China)	5
São Paulo (Brazil)	72
Buenos Aires (Argentina)	66
Maputo (Mozambique)	65
Johannesburg (South Africa)	57
Phnom Penh (Cambodia)	48

Source: van Dijk *et al.* (2007)

(Pansters and Berthier 2007: 36), causing people to restrict their movement, avoid leaving home at night and retreat into private spaces (see Table 6.1). Fear has always played a part in urban planning, from medieval cities with their walls and moats, through to Baron Haussman's complete renovation of nineteenth-century Paris' layout, with its wide boulevards designed to facilitate military control of the city (Lemanski 2004). However, such state-led strategies to mitigate fear have given way more recently to private and individual responses in the form of fortress-like gated communities. When viewed historically, this private retreat into walled enclaves since the early 1980s is one of the most significant spatial trends of urban development.

Gated communities predominate where inequalities are pronounced and public security is inadequate (see Box 6.3). Most commonly associated with North American cities (M. Davis 1990, 1998), the phenomenon is widespread in countries such as Brazil and South Africa where socio-economic inequalities are particularly dramatic (Beall *et al.* 2002; Caldeira 2000; Lemanski 2004). In South Africa alone, the number of private security guards has increased by 150 per cent since the late 1990s. In some cases, residents have fenced entire neighbourhoods, with gated access being monitored by armed private security guards, making it necessary for metropolitan authorities to legislate against the privatisation of public space. Even some public police stations are protected by private security companies (Human Security – Cities 2007: 27).

While walled enclaves and private security arrangements are generally confined to wealthy elites, there are many examples of low-income

Box 6.3

Africa's oil tycoons: multinational companies as gated communities

When Angola achieved independence from Portugal in 1975, the country was plunged into a civil war that lasted 27 years. The conflict was largely fuelled by the country's abundant natural resources – including oil. Multinational firms maintained their operations throughout the conflict, housing their employees in militarised urban spaces around the capital city Luanda. Once considered the 'Paris of Africa', Luanda was built for a population of half a million people but is today home to 4 million living in broken-down colonial villas and sprawling shanty-towns. The streets are strewn with rubbish and sewage and public provision of basic services is almost non-existent. As a result, the multinational oil companies pumping almost 1 million barrels a day from Angola's lucrative oil reserves generally opt to house their employees in self-sufficient suburban-style walled communities outside the city.

A campus of ranch houses built in the 1960s to house workers for the oil company now called ChevronTexaco remains in operation today. This self-contained compound known as Malongo provides many of the amenities of urban life including sports facilities, restaurants and luxurious bars. Its inhabitants live in almost complete isolation from the rest of the country, and the contrast with Luanda could hardly be starker:

> ChevronTexaco's own well and private filtration system supplies drinkable tap water – a rare luxury in Africa. Spacious dining halls offer a stunning array of fresh seafood, imported meats, salad and dessert bars. The vegetables are all grown in an organic greenhouse on the compound, set up by Norwegians, and bright green Granny Smith apples are flown in from South Africa. For entertainment, there's baseball, basketball, volleyball and tennis, a cricket pitch, horseshoes and a rolling green golf course. Unlike the rest of Angola, where the official language is Portuguese, the language of Malongo is English. If workers still get homesick, they can dial direct from their rooms to the United States – no need for international dialing codes. Indeed, except for the extraordinary bats that hang in ominous clusters from the branches of the compound's mango trees, you'd never know you were in Africa.
>
> (Eviatar 2004: 3)

Malongo is guarded and protected as rigorously as any military camp; no one can come or go without permission, the gates are guarded round the clock and the compound is double-fenced with barbed wire, between which is a ring of anti-personnel land mines. These mines were planted by the Communist government during the civil war, which granted the company drilling rights and agreed to protect Malongo from attack by rebel forces.

Source: Eviatar (2004)

settlements becoming private, fortified spaces as well. For example, in Johannesburg, some former hostels built to house migrant labourers working on mines and in manufacturing enterprises now operate as low-income private gated communities. Initially formed by ethnic minorities for protection, these 'fortified' hostels now in some cases cloak shady and illicit activities. As a result, the hostels are closed to the police and emergency services, especially at night, which works against the interests of many of their residents, especially the most vulnerable such as pregnant women and sick children (Beall *et al.* 2002). What distinguishes the gated communities of the rich from the enclaves of the poor is that in the latter residents are often obliged to choose safety and security over services. When people fortify themselves in closed compounds and exclude themselves from life in the city, a geography of fear ensues based on insider–outsider exclusions. People provide and maintain their own infrastructure and services and become disconnected from their surroundings, eschewing public space and opting out of local governance. Walled homes and fortress settlements are a long way off from Jane Jacobs' (1961) notion that busy public places offer natural surveillance and better safety due to more 'eyes on the street' and render rhetorical Peter Marcuse's question: 'Do walls in the city provide security – or do they create fear?' (Marcuse 1997: 101).

Fear coupled with a desire to establish a safe refuge from violence often results in voluntary segregation along ethnic, religious and racial lines, resulting in further mistrust and antagonisms between groups struggling to get by in a complex environment – sometimes with disastrous consequence. For example, in South Asia colonial urban planning under the British Raj gave rise to separate 'colonies' for different categories of state employee. These included elite settlements for different classes of civil servant and for the military, as well as low-income settlements set up for minority and low-paid workers such as the 'sweeper colonies' that housed municipal garbage collectors and sewerage workers. These colonially planned communal enclaves, often overlaid by class, ethnic or caste identities, persist to this day, and many effectively operate as gated communities. When political tensions reach a boiling point, residents in these communally drawn neighbourhoods sometimes come to blows. In Indian cities in particular, spatial segregation and communal violence between Hindus and Muslims living in such enclaves are widespread, being particularly evident in Mumbai and the Gujarati city of Ahmedabad (Chandhoke *et al.* 2008; Varshney 2002). In 2002, communal violence between Hindus and Muslims in the city of Ahmedabad resulted in the deaths of over a thousand people.

When cities gain a reputation for endemic crime and violence – whether it is socially, politically or economically motivated – their ability to generate wealth is undermined. As we discussed in Chapter 3, cities are dynamic economic spaces, but only if the benefits of agglomeration outweigh the costs. Crime and endemic violence represent significant costs and serve to encourage potential investors and skilled and educated workers – who can take their assets elsewhere – to move away, robbing such cities of the talent needed to compete in the global economy (UN-Habitat 2007). For those who do not have the option to leave, the costs of doing business in an insecure environment limit their scope for growth. The need to spend more revenue on private security and insurance diverts resources away from investing in the future. It is difficult to quantify the effects of crime and violence on economic development, but the available evidence indicates that the effects are significant. In Brazil, for example, it is estimated that some 10 per cent of annual GDP is spent on private security and insurance (Brennan-Galvin 2002), and the collapse of the country's public health system in the 1980s and 1990s has been attributed to the high costs associated with managing the sheer volume of homicides and crime-related injuries (UN-Habitat 2007: 46).

While lawless cities may discourage legitimate investors and consume resources that could otherwise be spent in pursuit of development, they frequently attract illicit operators eager to take advantage of poorly policed and regulated urban spaces. Indeed, the global criminal economy 'is solidly rooted in the urban fabric, providing jobs, income, and social organisation to a criminal culture, which deeply affects the lives of low-income communities and of the city at large' (Castells 2000: 395). Transnational criminal networks gravitate towards cities by virtue of their infrastructure and the cover they provide through the density of the built environment. They often penetrate the urban social fabric and affect the smooth operation of urban governance. These organisations thrive in the cities of low- and middle-income countries, with research suggesting that cities in Africa, Central Asia and Latin America provide particularly fertile ground for their activities (UN-Habitat 2007: 45). The problem is particularly pronounced in countries affected by conflict, where lawless cities serve as basing points for international criminal networks. Indeed, it is often through these conflict-affected cities that criminal networks most effectively integrate themselves into the global economy (Esser 2004). For instance, the small city of Bujumbura in Burundi has an estimated 200–300,000 small arms in circulation – a consequence of a vibrant regional arms trade that fuels ongoing conflict in the Great Lakes Region of Africa (Van Brabant 2007: 20).

Figure 6.1 The vicious cycle of public security failure

Source: Adapted from Human Security – Cities (2007: 26)

From the local to global level, efforts to tackle crime and endemic violence in cities are complicated by a feedback cycle that renders crime-ridden and violent cities prone to crime and violence (see Figure 6.1). Failed public security, weak urban governance, unequal access to economic opportunity for certain parts of urban society, and the naturalisation of a state of fear and insecurity within a city can become self-reinforcing phenomena (Agostini *et al.* 2007).

Breaking this cycle requires an understanding of the root causes of crime and violence, which include demographic, social and political factors. Rapid urbanisation in the face of poor urban planning and management generates the myriad insecurities discussed in previous chapters, including the formation of dense, poorly serviced low-income settlements. These large, precarious settlements are often interspersed with the gated communities and securitised spaces of rich urban dwellers. While crime is often associated with poverty, recent evidence suggests that the root cause of crime and violence is more likely to be inequality or social exclusion – thrown into stark relief under conditions of rapid urban growth – than poverty per se (Fajnzylber *et al.* 2002; Human Security – Cities 2007; Moser and Rodgers 2005; UN-Habitat 2007). Furthermore, cities are naturally heterogeneous spaces with people from many socio-economic and cultural backgrounds living and traversing a common urban space. In some cases, cultural or socio-economic tensions with roots outside the city are transferred to the urban context. In Pakistan, for example, peasants excluded from essential environmental resources by landholding elites in rural areas often migrate to the cities where new forms of competition over scarce urban resources ensue and inequalities are painfully evident, further fuelling existing communal and class-based rivalries (Moser and Rodgers 2005: 21).

Apart from the rapid growth of cities, another demographic phenomenon is often cited as contributing to urban security threats. Poor urban populations tend to be young (a consequence of the demographic transition discussed in Chapter 2), and youthful populations have long been linked to increased levels of crime and violence, especially where unemployment is rife. Unemployed youth represent a particular threat in the context of a public security vacuum, which can result from the breakdown of conventional social structures and/or the ineffectiveness of formal institutions of governance to contain or manage social discord. The situation is most acute in low-income areas, which are often virtually abandoned by public security services (Méndez *et al.* 1999; Pinheiro 1996). In the same way that informal water vendors step in to provide water in unserviced slums, in the absence of publicly provided security, other social forms – such as vigilante groups and gangs – emerge to fill the void. For example, in the slums of Managua, Nicaragua high levels of crime and insecurity in the 1990s were in some ways mitigated by the activities of youth gangs, which turned neighbourhoods into 'no-go' areas, protecting insiders from external harm by way of a ritualised form of warfare with outsider gangs. In such a context, gangs – which are more frequently associated with criminal activity – served a social function, 'a form of local-level social structuration in the face of socio-political breakdown' (Rodgers 2006: 269). However, such local level institutions can easily be perverted. When informal institutions fill a law and order void vacated by the state, they are usually left to operate with impunity, often with violent outcomes and with violence breeding further violence (Moser and Rodgers 2005: 24). In some cases, gangs 'become so powerful that they are able to successfully battle police for control over urban space' (D. Davis 2007: 19), thereby making the work of providing law and order all the more difficult for public authorities. In 2001, for example, almost half of the cities in Latin America and the Caribbean had areas considered dangerous or inaccessible to the police (Human Security – Cities 2007: 26).

There is also evidence to suggest that countries undergoing significant political transitions – from war to peace, and from non-democratic to democratic rule – often experience a rise in urban crime and violence (UN-Habitat 2007). The transition from war to peace can be turbulent and complicated by the demands of reintegrating former combatants. With few practical skills and minimal assets, former combatants often find few prospects for generating livelihoods beyond the private security sector or the shadowy interstices of the informal economy, creating ground fertile 'for the establishment of drugs, crime, and terrorist syndicates from which no country may be immune' (Caplan 2005: 256). The difficulty of transition from war to peace has been particularly evident in Central America, where 'crime is now so prevalent that levels of violence are comparable to, or higher than, those obtaining during the decade of war that affected most of the region during the 1980s' (Rodgers 2006: 268). In El Salvador, for instance, the annual number of deaths due to crime during the 1990s exceeded the average due to the war during the 1980s by over 40 per cent (Pearce 1998: 590). The transition from war to peace in Latin American countries was often linked to a transition to democracy – a process also associated with a spike in urban crime and violence, and homicide rates in particular (UN-Habitat 2007: 71). The restructuring of power relationships involved in such a transition sometimes inspires distorted interpretations of justice. This is illustrated starkly in South Africa, where many of those engaged in armed robbery and car hijacking following the first democratic elections in 1994 justified their actions in terms of wealth redistribution. On a more positive note, as a country moves towards the full institutionalisation of democracy, violence tends to recede again (UN-Habitat 2007: 71), offering some hope that inclusive politics can contribute to creating peaceful cities.

Criminal activities and urban violence proliferate in the absence of adequate public security. While demographic forces and political transitions are clearly beyond the control of urban authorities, a sound public security force can help to mitigate crime and violence by managing conflicts in a dispassionate manner. However, responsibility for public security does not lie exclusively at the city level. Indeed, many metropolitan governments and municipalities do not have their own police force and are reliant on protective and correctional services being supplied by national governments. Unfortunately, this is not always forthcoming. Strengthening the capacity of cities to provide public security is an important – and frequently overlooked – development priority.

Cities, war and terrorism

Warfare, although less pervasive and less common than crime and endemic violence, represents the most dramatic and destructive threat to human security in cities. By definition, war is an armed conflict between rival groups of people (e.g. nations, citizens, political parties). As such, many kinds of conflict can be classified as war. We focus here on two that exemplify the broader historical shift away from conventional wars (involving pitched battles between rival states and codified practices of engagement) and towards new modes of armed contestation that have important implications for urban security: 'slum wars' and 'new wars'.

Conventionally, war in developing countries has been thought of as a rural phenomenon. Names such as Amilcar Cabral, Mao Ze Dong, Ho Chi Minh or Ché Guevara all conjure up images of peasant guerrillas involved in anti-colonial or nationalist wars fought mainly in the countryside. However, the rural rebellions of the twentieth century have increasingly given way to armed conflict in cities (Beall 2007). This shift does not merely imply that conflicts once fought in the countryside are now taking place in urban spaces (although this is true), but also that the manifestations of armed conflict are evolving in an urban context. What Dennis Rodgers (2007a) refers to as 'slum wars of the 21st century' blur the line between organised rebellion and endemic violence, pitting groups of urban residents against each other and sometimes against state authority. Contestation is not necessarily violent or destructive: for example, it can involve the struggles of marginalised communities or groups against the state or urban elites, such as the Movimento Sem Teto in São Paulo, Brazil, which mobilises poor urban communities to fight for housing rights (see Box 7.4). However, contestation can manifest in more dystopian forms of collective action, such as urban gangs and vigilante groups that abide by laws of their choosing. The scale of violence that such 'social' forms of contestation generate is sometimes on a par with more conventional wars. For example, between 1978 and 2000 some 39,000 people lost their lives due to a protracted civil conflict between guerilla forces and the government of Colombia. Over the same period, over 49,000 people (particularly children) lost their lives to armed conflict in the slums of Rio de Janeiro, Brazil (Dowdney 2004). By the standard definition of civil war – i.e. an armed conflict resulting in the deaths of 2000 people or more – the slums of Brazil have been mired in civil war for decades. Clearly, the line between 'endemic violence' and 'war' is highly subjective, depending on what kinds of violence one chooses to use as a criteria for classification.

There is, however, a flip side to 'slum wars' of this kind: the violence exercised by states against their own citizens. This is more common than one might imagine, and can take several forms. In Karachi, for example, opposition groups such as the Pakistan People's Party (PPP) accused the military government in 2007 of fomenting some of the worst political violence this unstable city has seen in two decades. 'Shoot to kill' policies serve to exacerbate the situation while the exploitation of intergroup tensions in the city has turned political conflict into ethnic violence. A senior official of the PPP commented that 'People in civilian clothes were running around targeting anyone. They were firing from the tops of buildings and were even using tear gas canisters as the police stood idly by. The government is colluding 100 per cent' (quoted in *The Independent*, 15 May 2007).

However, state-enacted violence against urban residents can take a more subtle, but equally pernicious form: urbicide. Originally employed as a descriptor for deliberate attempts to deprive people of the benefits of urban life (Berman 1987), urbicide can include instances of pernicious urban planning, forced evictions, involuntary resettlement and 'the deliberate destruction of urban infrastructures for political purposes' (Beall 2006: 111). One example is Operation Murambatsvina in Zimbabwe, which translates literally from Shona as 'drive out the rubbish' or, more euphemistically, 'restore order' (Potts 2006). Part of the intention was to eradicate illegal (read: informal) activities in the capital city of Harare, with street vendors and others operating in the informal economy arrested and their businesses destroyed. Within six weeks an estimated 700,000 urban residents lost their homes and livelihoods, with up to 2.4 million people affected overall. Overnight, self-help housing and informal structures were declared illegitimate as bulldozers and demolition squads run by youth militia led the assault, which resulted in injury and even death. Families were forcibly removed to rural areas or peri-urban holding camps (UN-Habitat 2005). Operation Murambatsvina was a pre-emptive strike against an urban-based political opposition by way of a highly militarised national level operation within the context of Zimbabwe's severe economic and political crisis (Potts 2006). There were also continuities between colonial and post-independence attitudes to order and urban planning, with the eradication of informal economic activities and informal housing being pursued as zealously as before independence. However, when the assault came, neither formal nor informal dwellings were spared, with whole settlements being levelled to the ground (see Plates 7 and 8). Under the

Plate 7 An informal settlement in Zimbabwe before Operation Murambatsvina

Source: DigitalGlobe

Plate 8 An informal settlement in Zimbabwe after Operation Murambatsvina

Source: DigitalGlobe

rubric of public policy, the state of Zimbabwe committed a devastating act of urbicide.

More recently the idea of urbicide has been used to describe acts of terrorist violence and outright war (Graham 2003). In these contexts urbicide refers to the destruction of buildings and places, but also much more. When the Israeli Defence Force bulldozed a 40,000 square-metre area in the centre of the Jenin refugee camp in Northern West Bank, some 52 Palestinians were killed in the attack, half of them civilians, with several crushed to death in their homes. Operation Defensive Shield left 140 multi-family housing blocks completely destroyed, 1500 damaged and some 4000 residents homeless out of a total population of 14,000 (Graham 2003). Referring to Palestine, Stephen Graham argues:

> urbicide by bulldozer is not just about the demolition of homes and urban living spaces. House demolitions have long been paralleled by intensive infrastructural destruction, as Israeli forces work to prevent or systematically undermine the modernization of Palestinian urban society and the development of economic, technological, cultural or bureaucratic institutions.
>
> (Graham 2004c: 199)

From the 1980s onwards, the Israeli government systematically used the establishment of settlements to prevent or disrupt the consolidation of Palestinian urban centres. As such, urbicide reveals Israel's 'denial of the inevitability and necessity of Palestinian urbanization' (Graham 2004c: 209). Similarly in Bosnia, the war not only targeted the destruction of buildings symbolic of Bosnian culture but also constituted 'a relentless assault' on the urban fabric in which people carried out their social and collective lives: mosques, churches, synagogues, markets, museums, libraries and cafés (Coward 2004). This idea of destroying historic urban centres to gain political leverage is not, of course, new. In the Vietnam War for example, the bombardment of Hanoi was partly due to its symbolic significance as a cultural centre in a war that was actually being fought predominantly in rural areas. However, more recent international and civil wars have been fought largely in and around cities themselves. As such, many contemporary forms of urbicide are not only attacks on cities as cultural symbols but also attempts to flatten the complex urban landscapes that enemy combatants are using to their tactical advantage.

The use of urban space for tactical advantage is a particular feature of 'new wars'. New wars are said to be different from old wars in their causes, tactics and financing, with a number of these differences being linked to global economic processes (Kaldor 2006). Unlike ‘old wars’ that were fought according to respected rules of engagement designed to protect non-combatants, ‘new wars’ are characterised by the intention of generating fear, the effect of destabilising social and economic structures, and above all displacing people. As with contemporary urbicide, civilians are often targets in new wars, and cities provide them in abundance. Often it is particular groups that are targeted, as with the ‘ethnic cleansing’ that characterised the civil war in Bosnia-Herzegovina – in which the cities of Sarajevo and Mostar served as primary targets – and the civil war in Rwanda. Even when the main fighting occurs elsewhere in a country, cities are affected by people who have become refugees or internally displaced and who contribute to ‘the influx of villagers to towns and cities’ (Tibaijuka 2000). In Angola, for example, the capital city of Luanda grew in size fivefold during the war, while Mogadishu provides a particularly stark example of the dramatic effect that urban conflict can have on the demography of cities and the surrounding region. Between the end of January and middle of May 2007, 400,000 people fled Mogadishu and a further 300,000 were displaced within the city (*The Independent*, 15 May 2007). Yet policy makers remain largely unaware or unresponsive to the impact of war and conflict on cities.

In the aftermath of the US-led coalition’s invation of Afghanistan in 2001, the United Nations High Commission for Refugees (UNHCR), and associated agencies and NGOs, ensured that all returning refugees and internally displaced persons (IDPs) were sent back to their villages of origin. Many rural settlements had been devastated by war or drought, so that reconstructing livelihoods in the Afghan countryside was difficult. So, while the government and international donors poured money into rural support, people flocked from the countryside to the cities in search of urban livelihoods and alternative locations in which to build peaceful and secure lives (Beall and Esser 2005; Beall and Schütte 2006). It is when ordinary people become victims and when states fail to uphold their obligations to protect civilians that both formal and informal non-state organisations move in to fulfil this role. It is also often the trigger for humanitarian aid and the justification for international military intervention. In new wars, the monopoly of violence formerly held by the state gets taken over by local warlords, mercenaries, organised criminal groups, private militias and so on, and their operation and survival is closely associated with the emergence of so-called ‘war economies’ (Collier 2000). Here too cities are vital hubs that connect the violent economies of crisis states into the transnational networks of which they often form a part. Even in Afghanistan, where ongoing hostilities are fuelled and financed by the involvement of the Taliban and local warlords in poppy production for opium (a largely rural pursuit), cities and particularly border cities act as important conduits into international markets.

New wars also cannot be understood independently of regional and global forces, and the growing importance of cities in insurgency operations and asymmetrical warfare. This is not to suggest that wars in previous eras were free of international links and influences, but rather to highlight the way in which cities are increasingly affected. As Graham has argued:

> Just as it is no longer adequate to theorize cities as local, bounded sites that are separated off from the rest of the world, so, similarly, political violence is now fueled and sustained by transnational networks that can be global and local at the same time.
>
> (Graham 2004a: 3)

Cities moved most dramatically to centre stage with the collapse of New York's World Trade Center on 11 September 2001, illustrating beyond doubt both the susceptibility of cities to war and terror as well as the impact to be gained from 'spectacular violence' (Goldstein 2004). Two and a half years later this was confirmed when bombs were detonated on packed commuter trains in Madrid, killing 191 and injuring over 1500 people, and again in July 2005 with the London bombings, which left 52 people dead and over 700 injured. These events signify the 'implosion of global and national conflicts into the urban world' (Appadurai 1996: 152–153) and cities of the Global South do not escape this trend. Cities from Karachi and Kabul to Mumbai, Nairobi and Bogotá, have all been targets for episodic or sustained acts of terror.

Terrorism is not a new phenomenon, and in fact the annual number of terrorist attacks worldwide has been falling since the end of the Cold War as state support for nationalist and ideological struggles faded (Sandler and Enders 1999). However, the downward trend in terrorist incidents has been accompanied by a rising trend in the number of casualties (defined as deaths and injuries) per incident (see Figure 6.2). In other words, on average there are fewer attacks every year today than there were a decade ago, but attacks have become more deadly. This shift is due in part to a steady rise in religious extremism since the late 1980s, as well as a rise in the number of non-state affiliated groups who view civilians as legitimate targets (Sandler and Enders 1999). Although not exclusively an urban phenomenon, the overwhelming majority of terrorist attacks target cities

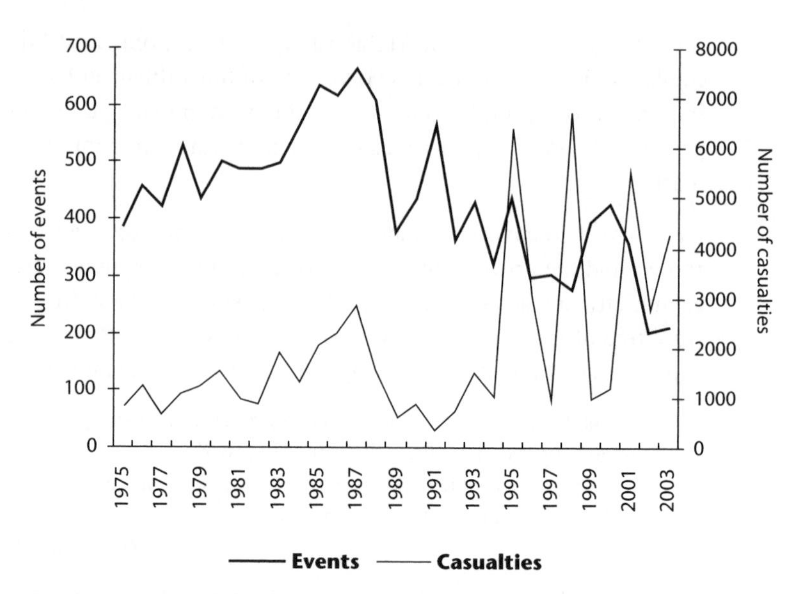

Figure 6.2 Transnational terrorism: incidents and casualties, 1975–2003

Source: Sandler and Enders (2005)

due to the high concentration of civilians and strategic infrastructure they provide.

As well as serving as targets, cities function as nodes for the articulation of international terrorist networks. In the 2005 bombings in London, the attackers were British with links to Pakistan, their country of origin. In a 2007 incident, the suspects were Iraqi medics working in Britain. In both cases their motivation was support for causes rooted in the Arab world, such that grievances felt in one part of the globe can be felt in many others:

> The geography of terror has moved on to the global stage largely by way of cities and specific urban symbols. In part the message of the London bombers was that any war perceived to have its origins in London would come back to roost in London. While London was the site of thousands of people of multiple origins, taking to the streets in protest against the war, 7th July 2005 also stands as a symbolic gesture against the Anglo-American alliance that went to war in Afghanistan and Iraq. In a quintessentially cosmopolitan city like London, the events of July 2005 are further emblematic of how the victims of the bombs were inextricably linked to families, friends and sympathisers across the globe.
>
> (Beall 2006: 107)

It is important to recognise that an attack on a given city is not necessarily an attack on the city itself. For instance, the bombing of busy urban tourist centres in Bali and Sharm el Sheikh were aimed not at the towns but rather to communicate messages across the globe. In a sense, urban terror constitutes an international language; a means of projecting political messages across the global media landscape. However, the impacts of terrorist attacks on cities in low- and middle-income countries are particularly devastating due to the limited resources and capacities these countries have to devote to recovery. For example, the al-Qa'ida bombing of a hotel in Mombassa in November 2002 was to the residents of that city the last straw in the region's downward spiral into poverty and banditry (Richards 2002). Moreover, in the context of weak governments and struggling economies, such as in Afghanistan, it is even more difficult to undertake reconstruction, both physical and social (Beall and Esser 2005).

It is also important to recognise that cities are as vulnerable to the 'War on Terror' as they are to terrorism, caught as they are in the crossfire of international conflicts. As Peter Marcuse has argued: 'Not terrorism but what has been done under the mantel [sic] of counter-terrorism, has had a significant effect on cities since the attack on the World Trade Center on September 11, 2001' (Marcuse 2004: 263). Urban warfare is the most destructive, manpower intensive and complex of all forms of engagement, and therefore from a military point of view, it is something that has been avoided at all cost. However, there is now ample evidence to show that conventional warfare is coming to terms with the inevitability of urban combat despite the fact that cities constitute a difficult terrain in which to conduct war. Even when 'collateral damage' in the form of non-combatant casualties is avoided, civilians 'quite simply get in the way' (Ashworth 1991: 114). For example, the occupying forces in Afghanistan rejected area bombardment in favour of precision attacks on elements of urban infrastructure that could support Taliban resistance, in the hope of avoiding civilian casualties. However, most targets in the capital city Kabul were surrounded by slums, with most civilian deaths occurring in these densely populated areas:

> Cumulatively, these numerous small death tolls meant that the US-led bombing campaign, however well-designed to minimize civilian casualties, was the most lethal in terms of bomb tonnage since the Vietnam era, resulting in between 2,214 and 1,571 civilian casualties.
>
> (Esser 2007: 14–15)

Technology works best in conflicts with known features rather than in urban guerrilla warfare situations. The experience of the Iraq War has shown that military hardware is not enough. Alongside high-tech air bombardment, US troops went in with wire-cutters, ladders and mirrors to help them see around corners, and had to engage in very basic forms of street fighting at ranges of 6 and 60 rather than 60,000 feet. With or without technology, urban warfare is difficult because cities are rarely empty and 'intentionally or accidentally, non-combatants shape the battlespace' (Hills 2004: 238).

Reconstruction, peace-building and development

When hostilities cease, whether they have manifested as urbicide, regular terrorist incidents or outright warfare, the challenges and opportunities of reconstruction begin. In the aftermath of war, in particular, the financial and technical challenges of repairing the physical fabric of a city can be overwhelming. Basic infrastructure, such as roads, water, sewerage and power supplies are often left in a dysfunctional state, while buildings might be destroyed completely or in part. Heritage areas and sites of historical value need to be rehabilitated and preserved. The violent assault on Baghdad's infrastructure by US forces during the first Gulf War in 1990–1991 nearly reduced what was previously a fairly advanced economy to a 'pre-industrial age' (Banarji 1997: 199). Following the second assault on Baghdad in the more recent 'War on Terror', the combination of economic and military violence once again 'reduced the Iraqi economy . . . to conditions that rival those of the most degraded nations in the Global South' (Schwartz 2007: 21). The rebuilding process also offers unprecedented opportunities for profit.

Transnational corporations often step in to execute 'aid' contracts worth millions – sometimes billions – of dollars, raising concerns that 'reconstruction' diverts large volumes of aid resources otherwise destined for poverty alleviation and development (Beall *et al.* 2006). At a global level, huge resources have recently been directed towards counter-terrorism measures and channelled into surveillance, emergency planning and capacity building, with detrimental effects for development. In both Tanzania and Kenya, for example, USAID has invested in training centres and emergency planning regarding terrorism, when these countries still lack the fundamental infrastructure to deal with hazards such as floods, droughts and public health issues.

There is also some concern that reconstruction efforts overlook the deeper – and less visible – legacies of conflict. Hostilities disrupt the rhythms and practices of daily life by destroying the assets of ordinary citizens, preventing children from attending school, encouraging the emigration of those with the means to do so (often the most wealthy and educated) and internal displacement of those seeking refuge. In contexts where ethnic or communal rivalries are involved, a peace settlement does not resolve the memories of the atrocities of each side against the other. In other words, cities emerging from conflict are confronted with the challenges of rebuilding a shattered society in physically and socially circumscribed spaces. Hence reconstruction is not merely a question of bricks and mortar; it requires identifying and confronting the social and psychological inheritances of a conflict.

The very features that make cities prone to conflicts and violence of all kinds – i.e. their size, density, diversity and concentrations of wealth and power – also make cities promising spaces for reconciliation and reconstruction. In many cases, cities are 'microcosms of broader social fault-lines and tensions affecting a nation' (Bollens 2006: 67). Where societal divisions are inscribed in the fabric of a city, the dividing lines themselves are often the first sites of reconstruction through deconstruction: the Berlin Wall was brought down in a dramatic rejection of Cold War polarities; South Africans wilfully defied (legislated) residential segregation in cities; and residents of Nicosia, Cyprus regularly crossed the 'green line' separating the Greek and Cypriot sides of the city. These dividing lines, which often seem to be 'intractable urban fixtures', can be transformed into symbolic spaces of reconciliation and unity, performing 'important roles in stitching together torn cities' (Bollens 2001: 187). More generally,

> The city is important in peace building because it is in the streets and neighborhoods of urban agglomerations that there is the negotiation over, and clarification of, abstract concepts such as democracy, fairness and tolerance . . . Peace building in cities seeks not the well-publicized handshakes of national political elites, but rather the more mundane, but ultimately more meaningful handshakes and smiles of ethnically diverse urban neighbors.
>
> (Bollens 2006: 67)

Urban planners, in particular, can play an important role in the aftermath of crisis by translating seemingly abstract and insoluble conflicts into (literally) concrete, tractable public projects that 'constitute leading implementation edges of new democratic goals and mechanism' (Bollens

2006: 75). But the process of planning and implementing reconstruction strategies in cities requires more than optimistic planners and a vision of peace. Negotiating initiatives and priorities must involve all stakeholders, from the community, city, national and international arenas. In Afghanistan, for instance, the reconstruction of Kabul has been complicated by poor coordination between various stakeholders, and numerous property disputes arising from war and regime change, which have resulted in protracted land tenure disputes and in some cases evictions and land seizures. These have been particularly acute given the rapid growth of the city population following the US invasion in 2001 and the subsequent skyrocketing of land prices in the city (see Box 6.4).

Box 6.4

Reconstructing Kabul

The challenges of post-war reconstruction are well illustrated by Kabul, capital of Afghanistan. Badly but not uniformly destroyed by a series of wars – most recently the US-led coalition invasion of 2001 – the city has been struggling to reconcile the dramatic changes that conflict has imposed upon it. Those with the means and networks to do so fled Kabul, while others from every corner of the country sought refuge in the city, particularly after comparative peace was restored. Between 2001 and 2004 Kabul's population grew from half a million to 3 million people, despite the decrepit state of the city's infrastructure, soaring land and housing prices and high levels of unemployment.

The Afghanistan government's National Development Framework identified the restoration and extension of urban infrastructure as a priority, although international donors have tended to place more emphasis on rural rehabilitation and development. Those agencies that have committed resources to urban reconstruction have made halting progress due to poor coordination. For example, UN-Habitat and a number of NGOs provided or augmented basic infrastructure and services such as the provision of wells and waste collection schemes, often without reference to efforts to restore and protect Kabul's fragile water system. Government agencies also lacked a coordinated approach, with tensions between the lead urban agency, the Ministry of Urban Development and Housing (MUDH) on the one hand and the Mayor and Kabul Municipality and Kabul City High Commission on the other. The result was that urban management was fragmented across several levels of government and ministries, including the Ministries of Transport, Public Works and Water and Energy, as well as other organisations. In the mean time, people were getting on and reconstructing their own homes and neighbourhoods, ignoring the plans and forums of official agencies.

In response the Urban Sector Consultative Group (USCG) was formed to bring together these government organisations, international donors and financial institutions and some local and international NGOs working in the urban sector and to assist the MUDH with policy and programme development and implementation. In addition efforts are made to consult and coordinate with Afghan and other contracting firms and developers, as well as community groups (shuras) and headmen (wakils) across the 470 or so neighbourhoods. As such the USCG provided an important start towards sharing information, building ownership and coordinating activities among the multiple actors involved. While international agencies have given valuable support, in the long run the residents of Kabul and the Afghan government will have to sustain the reconstruction effort, and success will depend upon continued coordination, collaboration and engagement.

Source: Beall and Esser (2005)

Whether or not urban reconstruction efforts succeed depends upon the willingness and capacity of the various stakeholders involved to seize the window of opportunity that a peace settlement provides. The aftermath of conflict inevitably entails a period of flux, in which new visions for society can be advanced or thwarted, new strategies for inclusive governance can be institutionalised or destabilised. Failure to seize this opportunity can result in a reversion to hostilities and the creation of new conflicts. This was painfully apparent in Iraq following the invasion of 2003, where the failure of authorities to provide law and order in the streets, consult with local community leaders and make concerted investment in restoring critical urban infrastructure precipitated a wave of sectarian violence.

Conclusion

Human security is a prerequisite for development. Insecurity is a direct threat to health and psychological well-being. But the effects of insecurity have longer-term implications. Without a minimal level of human security, people are not free to invest in their own (or their children's) future. Without a minimal level of security of person and property, economies cannot generate wealth. In countries in conflict, cities are sometimes the only areas that continue functioning. Those who advocate state-directed approaches to development take for granted that states are able to secure their territories from external threats, provide law and order in their streets and maintain a steady level of investment in the infrastructure and social services that are critical to enhancing people's basic capabilities, such as healthcare and education. Unfortunately, this is too often not the case, especially at local government level. Securing human security inevitably requires a 'bottom-up' approach. A well-trained, well-equipped police force is an important ingredient to ensuring human security, but strong communities and open forums for dialogue between groups with conflicting interests are equally necessary to prevent these conflicts of interests from devolving into open, armed hostilities. To effect these conditions requires political, policy and planning processes that are inclusive and responsive.

Unfortunately, cities everywhere are becoming increasingly fortified, militarised spaces. Perceived security threats are being translated into urban planning and governance strategies designed to control, survey and defend urban space:

> the design of buildings, the management of traffic, the physical planning of cities, migration policy, or the design of social policies for ethnically diverse cities and neighborhoods, are being brought within the widening umbrella of 'national security'.
>
> (Graham 2004a: 11)

This defensive mentality undermines the prospect for cities to be 'privileged spaces of democratic innovation' (Borja and Castells 1997: 246) by inscribing difference and divisions in the urban fabric. What is needed – now more than ever – is a focus on the ways in which contentious urban politics can be managed through progressive planning and inclusive governance strategies that accommodate contestation, cooperation and compromise. It is to these issues that we now turn.

Summary

- **In the increasingly 'globalised' post-Cold-War world the concept of 'human security' refers to a concern with mitigating the various forms of insecurity affecting ordinary citizens, irrespective of national boundaries and sovereignty.**
- **Violence can be conceptualised as politically, economically or socially motivated. In practice, however, these distinctions are often blurred and it can be more helpful to conceptualise how violence is manifested, from 'anomic' to more organised forms.**

- **For urban dwellers crime is the most pervasive source of violence, resulting in more deaths globally than war. Violence and crime also result in segregation, fear, capital flight and increased expenditure on private security, all of which have a pronounced negative impact on urban and national development.**
- **Periods of political transition to democratic rule or from war to peace can result in heightened violent crime, a problem exacerbated by the large numbers of former combatants and small arms that tend to gravitate towards cities in such periods.**
- **War itself is becoming an increasingly urban phenomenon, particularly in the context of the 'new wars', in which violence is aimed deliberately at civilians concentrated in urban space.**
- **Cities are also vulnerable to the increasing destructiveness of international terrorism due both to their density and symbolic importance. Acts of 'urbicide' by states or non-state actors disrupt urban infrastructures and disperse urban populations for political purposes.**
- **In the aftermath of war, cities hold potential as sites of peace-building and reconciliation, as well as offering great opportunity for profit. However, in the scramble for post-war reconstruction contracts conventional development projects can often be sidelined.**
- **In cities emerging from crisis, effective reconstruction – both physical and political – is dependent on good urban planning that is inclusive, responsive and supported at international and national as well as local and city levels.**

Discussion questions

1. Is it helpful to draw a distinction between human security and national security? Discuss with reference to the following:

   (a) international terrorism
   (b) civil conflict or 'new' wars.

2. 'Urban violence is either political, economic or social'. Discuss.
3. What are the various dimensions of the relationship between violent urban crime and development?
4. Why are so many states, armed groups and terrorist organisations motivated to commit acts of 'urbicide'?
5. How can the demands of post-conflict urban reconstruction best be managed in relation to longer-term development goals? What are the complementarities and trade-offs between the two?

Further reading

Beall, Jo (2006) 'Cities, terrorism and development', *Journal of International Development*, 18(1): 105–120.

Bollens, Scott (2006) 'Urban planning and peace building', *Progress in Planning*, 66: 67–139.

Graham, Stephen (ed.) *Cities, War and Terrorism: Towards an Urban Geopolitics*, Oxford: Blackwell.

Human Security – Cities (2007) *Human Security for an Urban Century: Local Challenges, Global Perspectives*, Ottawa: Foreign Affairs and International Trade Canada, March.

Koonings, Kees and Dirk Kruijt (eds) *Fractured Cities: Social Exclusion, Urban Violence and Contested Spaces in Latin America*, London: Zed.

Rodgers, Dennis (2006) 'Living in the shadow of death: gangs, violence, and social order in urban Nicaragua, 1996–2002', *Journal of Latin American Studies*, 38(2): 267–292.

UN-Habitat (2007) *Global Report on Human Settlements 2007: Enhancing Urban Safety Security*, London: Earthscan.

Useful websites

Cities and Fragile States component of the Crisis States Research Centre: www.crisisstates.com/research/cafs.htm

Conflict in Cities and the Contested State – a research project on divided cities: www.conflictincities.org

Human Security – Cities Project: www.humansecurity-cities.org

UN Habitat Safer Cities Programme: www.unhabitat.org/categories.asp?catid=375

7 Shaping city futures: urban planning, governance and politics

Introduction

In the face of rising urban poverty, insecurity and environmental change, what can or should be done to ensure a more equitable, inclusive and sustainable future for the world's population? There is no simple answer to this question. Every city, nation and region faces its own unique challenges, even if they share common problems such as rapid urban growth, increasing informality or climate change. Prescriptions are neither obvious nor desirable. However, in approaching this question, certain critical themes provide a useful starting point for context-specific analysis. In this concluding chapter we focus on the ways in which urban planning, governance and politics relate to the challenges and opportunities associated with urbanisation. Although we have touched on these issues throughout the book, we now bring them to the fore in order to highlight their paramount importance in understanding and effecting processes of urban change.

Building better city futures demands proactive urban planning that establishes a framework for urban growth. To say this, however, is not to suggest planning offers a clear remedy. Rather it implies a vision for urban development and processes of prioritisation. Who formulates and approves this vision? Who must sacrifice what in the process of achieving it? We address these questions in the second section of the chapter, looking specifically at the interaction between governments and citizens. It is this relationship that sits at the heart of important and ongoing debates about urban governance. Like 'effective urban planning', 'better urban governance' is often touted as a solution to urban problems. However, 'governance' is better deployed as an analytical tool than a tangible solution. Properly understood, an analysis of the obstacles to improving urban governance alerts us to the many actors, diverse processes and multiple scales that public action and urban development strategies must take into account in order to be effective and sustainable. However, 'good' urban governance is, like planning, often approached from a technical point of view when in fact it is deeply political. Effective coordination and decision-making among multiple stakeholders is often coloured by conflicting interests and visions. As a result, change is always political. In the final section we return to a theme highlighted in Chapter 1 that often informs urban politics – namely the importance of attitudes about urbanisation and urban development in determining the potential for progressive change. In particular, we reiterate the need to place urban development at the centre of the development agenda if we are to tackle the three greatest challenges facing the world in the twenty-first century: poverty, security and climate change.

Urban planning and policy

As noted in Chapter 2, urban planning – being the practice of designing and regulating urban spaces – dates back nearly to the birth of cities. However, as a discipline and profession, urban planning is a relatively recent endeavour. Prior to the demographic transition and industrial revolution in Europe, cities, once established, tended to have relatively static populations. It was only when urbanisation began to gather pace in nineteenth-century Britain (and subsequently elsewhere) that the need for more proactive, ongoing interventions was recognised. Indeed, according to planning scholar Peter Hall, 'twentieth-century city planning, as an intellectual and professional movement, essentially represents a reaction to the evils of the nineteenth-century city' (Hall 2002: 7). Burgeoning slums, outbreaks of infectious diseases, and gross inequality alarmed the ruling and middle classes in European cities, instilling a combination of guilt and fear of revolution. The response was a wave of research, legislation and action involving private individuals, government officials and philanthropic organisations to address the growing problems of urban poverty and inequality. In other words, there was a collective realisation that rapidly growing cities needed forward planning to tackle contemporary challenges, while also catering for the future. As one academic and veteran World Bank urban specialist has put it: 'In the city, *laissez-faire* means give up and suffer the consequences' (Cohen 2001: 57).

Many early urban planners – such as Ebenezer Howard, arguably the most influential figure in the intellectual history of urban planning – identified specific areas demanding professional attention including housing, infrastructure development, public transport and regulation of the built environment (e.g. in the form of building codes and land-use planning). However, Howard and others were often equally concerned with an almost radical agenda of social reform – a sense that urban planning could be used to create more equitable, healthy and productive communities (Hall 2002). Over the first half of the twentieth century, their ideas and visions were consolidated into an approach to planning that was highly centralised. New towns with adequate housing, ample open space and land-use patterns designed to separate domestic, commercial, recreational and productive life were designed and built. Large areas of cities designated as slums were razed to the ground and rebuilt from scratch. Importantly, the ideas of planning that emerged in response to the rise of Victorian slums reverberated around the world. The ideas developed in countries such as Britain, the United States, Germany and France were exported in the late colonial era to countries in Africa, Asia and the Americas. By the middle of the twentieth century a decidedly modernist vision of city planning had spread across the globe. In the process, many of the social concerns of urban planning pioneers had given way to top-down, universalising approaches that were overly rational and insensitive to social and cultural diversity (Allmendinger 2002).

In 1961, Jane Jacobs published a trenchant critique of the discipline and practice of urban planning, arguing that rigid visions of how the ideal city *ought* to work are invariably inconsistent with how they *actually* work. The fundamental problem with modernist urban planning, even when well intentioned, was that it was used to 'put people in their place' without paying much heed to what people were doing and what they wanted. Modernist urban planning simply defined appropriate spaces for particular groups and for designated activities. Common everywhere are zoning regulations that determine some areas as residential and others as commercial or industrial. Technical planning systems gave rise to static blueprints such as master plans that were *de rigeur* and rigidly applied in cities across the world. Planners adhered to master plans even if a city expanded in ways and directions that bore no resemblance to what was represented on paper or anticipated by the master plan. In the worst cases, modernist planning principles tinted with imperial ambitions and overt racism produced cities designed to segregate and control particular communities – as we saw in Chapter 2 in relation to Africa and Asia. Some critics even saw planning as a reformist tool that was complicit in the uneven and highly unequal outcomes of mainstream development processes (Escobar 1995).

The excesses and failures of modernist urban planning, which were as pronounced in low- and middle-income countries as in rich ones, inspired an intellectual movement advocating a more collaborative or communicative approach to planning. This shift reflected the realisation that people, through their everyday activities, also make cities, and often in ways that ignore the visions of planners. This is particularly apparent in the fast-growing cities of low- and middle-income countries where the reach of planning is limited: people create their own settlements informally, without a centralised vision or coordinating entity, and service them according to their own means. Hence informal settlements often do not conform to the grid pattern street layouts favoured by master plans. Instead, houses and lanes are arranged in ways that facilitate social interaction and communal and commercial activities. Similarly, livelihood activities often take place in the home or neighbourhood, in defiance of zoning legislation. Planning scholars became progressively aware that cities are 'more than the infrastructures, codes, and inputs necessary to manage population sizes and built environments' (Simone 2004: 240). Increasingly, the seemingly chaotic dynamics of urban spaces are celebrated as constructive and creative, in contrast to the dystopian view that uncontrolled urban areas are incubators of disease, vice and social discord.

This new perspective has contributed to a rejection of rationalist approaches in favour of planning as 'an interactive process, undertaken in a social context, rather than a purely technical process of design, analysis and management, rather than something handed down through normative frameworks' (Healey 1997: 65). The origins of modernist planning, which rested in ideals of social reform, have given way to concern with diversity and people's rights in the city. Planners are increasingly eschewing command and control functions in favour of greater consultation and participation and more flexibility in urban planning processes. The ubiquitous planning dilemmas of massive demand and limited resources, and of competing priorities and conflicting interests, are to be resolved through negotiation and debate. However, this shift has been slow and patchy, at best. For many urban planners trained in the modernist tradition, appeals for more collaborative and interactive approaches are difficult to countenance (see Box 7.1). Adopting collaborative or communicative planning implies a shift of power relations, giving greater

Box 7.1

The contemporary legacy of Kabul's 1978 master plan

The Kabul Master Plan was developed in the mid-1970s, well before the Soviet invasion of Afghanistan. It was approved in 1978 although implementation was constrained by over two decades of violent conflict and the fact that war gave rise to urban destruction, social fragmentation and displacement on an unprecedented scale. The Kabul Master Plan catered for a city of 700,000 people rather than the estimated 3.5 million currently living in the city. Although it has now been officially suspended through a presidential decree, the legacy of the Kabul Master Plan remains, with its maps and drawings still in circulation. Areas of the city covered by the plan receive attention and resources before those that are not, with the effect that infrastructure provision, such as a citywide water project, bypasses one of the most needy areas of the nation's capital city.
A survey of urban issues in Afghanistan states the following:

> The most critical challenge facing the urban sector in Afghanistan is to update current approaches to urban planning and management. In effect this

> means giving up the commonly held commitment both to existing master plans and the professional practice of master planning.
>
> (Beall and Esser 2005: 49)

However, there is a widely held view among urban professionals in Kabul that planning is no more than design and a reluctance to consider collaborative approaches or strategic urban development.

Sources: Beall and Esser (2005); photograph by Daniel Esser (2005)

voice to citizens and challenging the influence of entrenched political and economic interests – a subject to which we return below.

Despite this obstacle, there have been some notable successes. The most internationally celebrated form of collaborative planning at city level is participatory budgeting. It emerged in the Brazilian cities Porto Alegre, Belo Horizonte and Recife. Several factors conspired to facilitate it, including an increase in local government finances as a result of decentralisation reforms and the growing presence of leftist parties in local government. It is a process whereby local government officials work with neighbourhood groups and social movements to develop citizenship skills, increasing local accountability in a set of engagements involving ordinary people in the planning of the annual capital budget of a city. Nowadays more than 70 cities across Brazil have adopted participatory budgeting and have adapted the system to their needs, reflecting the ideals of a planning process 'where mutual understandings are forged through continuing collaboration' (Friedmann 2007: 997). It has spread to other countries in Latin America and has travelled to Europe, having been adopted by cities in Spain and Portugal and influencing a nationwide programme in the United Kingdom. This provides an interesting example of urban innovations from middle-income countries influencing policy and planning processes in fully industrialised nations (Beall 2005).

Participatory budgeting is not, however, without critics. They point out that the very poorest in society often do not have the time or awareness to participate and that there are persistent difficulties in deepening and broadening deliberative democracy. This raises an important point about the limits of participatory approaches to development: there are always restrictions with regards to who and how many people can effectively participate in decision-making processes. There are always local power dynamics that favour the views of some over others and participatory approaches can never ensure that everybody's views are adequately represented. Moreover, in the case of participatory budgeting in Brazil, it is not the whole local budget that is determined by the participatory budgeting process but rather only those aspects pertaining to infrastructure investment. This is usually somewhere between 5 and 20 per cent of the total budget. This has led some to question the real impact of the process on government spending (Souza 2001). Furthermore, the success of participatory budgeting in Porto Alegre and Belo Horizonte cannot be expected to translate easily and without adaptation to other cities in different contexts. Participatory budgeting has to take account of local circumstances and be carefully tailored towards them (Cabannes 2003).

Urban planning is, after all, an inherently normative and interventionist process. What distinguishes better and worse forms of urban planning is the ethical basis on which normative agendas are formed, who decides priorities and on whose behalf, and the processes by which decisions are made. As such, 'good' urban planning can be assessed only in relation to its fit with local context. In the past urban planners were celebrated for their (literally) concrete visions of the future town or metropolis involving maps, diagrams and models. Now, however, there is widespread recognition that the role of urban planners is to facilitate

> a communicative process in which formal plans, maps and other documents have a part to play but where the written word is less important than direct speech, and where mutual understandings are forged through continuing collaboration.
>
> (Friedmann 2007: 997)

Such an approach to planning is particularly relevant in the low- and middle-income countries that are currently experiencing rapid urban growth. In nineteenth-century Europe and North America, urban planning was born to tackle this challenge, but the planning strategies that emerged were ill suited to the task and poorly equipped to manage rapid change. The latter demands flexibility, and flexibility demands constant communication and collaboration, which not only enhance the probability of effective solutions, but also advance real freedoms by giving a greater number of people an opportunity to participate in deciding what their future will be like. In a sense, the primary objective of urban planning has shifted from one of articulating a concrete vision of an ideal city, to embracing an iterative process of negotiated action. To ensure inclusive, productive and sustainable city futures, there is an urgent need to strengthen and broaden the capacity of urban planners in those countries experiencing rapid urban change, as well as advocate flexible and inclusive planning processes.

However, better urban planning is only one pillar of an effective urban development strategy. In Chapters 3, 5 and 6 in particular, we highlighted the complex ways in which cities and towns are integrated into the regions of which they are a part – as well as their articulation with global political and economic forces. Many of the problems facing urban areas do not arise simply from within their boundaries, and what happens in cities has implications beyond the nebulous borders of any single urban settlement. As such urban planning has to be accompanied by urban policies that emanate from multiple spheres of government. Most often, a nation's 'urban policy' is not clearly articulated, but rather comprises a *pot pourri* of diverse policies and agendas. Strengthening the capacity of urban planners at municipal and metropolitan levels and advocating communicative planning practices are important objectives when confronting the challenges of urban development, but the real challenge is to achieve a coherent repertoire of urban policies that play themselves out at 'a range of scales but focused on cities' (Cochrane 2007: 141). The urban challenge requires, therefore, integrated efforts across all levels of government. With regard to urban poverty alleviation, this is well understood, for example, by the Swedish International Development Cooperation Agency (SIDA):

> Efforts to alleviate urban poverty must reflect a deep understanding of its causes and characteristics; a clear articulation of urban poverty dimensions in local development plans; and a comprehensive treatment of urban development in national plans and priorities to combat poverty. Urban development and poverty alleviation within urban areas also require an integrated effort across all levels of government to support efforts at local and national levels.
>
> (SIDA 2006: 7)

In a similar vein, John Friedmann (2007) argues more generally that effective urban planning cannot be conceptualised only at the urban scale, but rather 'will involve public interventions in housing, education, health services, transportation, sanitation, community and cultural affairs, and all the rest, in a "whole of government approach"' (Friedmann 2007: 996). According to Friedmann,

> Such an approach is possible only if there is broad agreement within local government on a long-term strategic vision. Building a local consensus is only a beginning, however. Beyond the local bureaucracy there are not only the bureaucracies of senior governments and

> powerful corporate interests that must be coaxed along; proceeding downwards to the lower echelons of the city's bureaucracy, the region's county and township governments, urban neighbourhoods and community-based organizations will all have to be consulted, because it is here that interventions must ultimately come to rest.
>
> (Friedmann 2007: 996–997)

This observation brings us to the broader question of how to cultivate effective urban governance, a concept that captures both the complexity of managing urban environments as well as the importance of the relationships between the diverse actors whose interests converge in urban spaces.

Urban governance, decentralisation and the metropolitan challenge

The word 'governance' has become ubiquitous in development discourse since the late 1980s, and yet the concept of governance remains somewhat fuzzy in many people's minds. One of the most often cited general definitions of governance comes from the United Nations Development Programme (1997):

> Governance can be seen as the exercise of economic, political and administrative authority to manage a country's affairs at all levels. It comprises the mechanisms, processes and institutions through which citizens and groups articulate their interests, exercise their legal rights, meet their obligations and mediate their differences.
>
> (UNDP 1997: 2–3)

There are essentially two important dimensions to this definition that need unpacking. The first is the scope of the term. The 'management of a country's affairs at all levels' suggests the involvement of a wide range of actors from national government agencies and ministries down to the smallest administrative units in a country. The second half of the definition introduces a range of non-governmental actors ('citizens and groups') and emphasises the ways in which these actors engage with one another – or the 'processes and institutions' that mediate these relationships. The concept can be distilled further as referring 'to questions of control over decision-making about how resources are used in a sea of competing and different interests' (Pieterse 2008: 5), or even more succinctly: 'the relationship between civil society and state, between rulers and ruled, the government and the governed' (McCarney *et al.* 1995: 95).

A recent trend in thinking about development – frequently referred to as the 'good governance' agenda – emphasises the need to build stronger, more accountable relationships between diverse actors in order to ensure that a society's resources are managed fairly and efficiently over time. The adoption of a discourse of governance reflects an acceptance of the failures of both the state-directed and market-led development paradigms, and the need for a more nuanced approach that recognises the potential positive synergies between states, markets and the wide range of actors that fall into the equally broad category of 'civil society'. It also reflects a realisation that traditional approaches to delivering development assistance – such as providing more resources (e.g. through aid), building government capacity (e.g. through training and technical assistance) or unleashing market forces (e.g. through extensive privatisation and deregulation) – do not get to the core of the problem.

It is important to be clear about the differences between 'good government' and 'good governance'. These differences can be illustrated by the new public management (NPM) approach to urban management (discussed in Chapter 5). The widespread promotion of NPM principles, which emphasised the need for non-state actors to take over many of the erstwhile responsibilities of governments (notably private companies and social sector organisations such as NGOs and community-based groups), was a response to the failure of states to run their cities and deliver services effectively. In cities, NPM was essentially introduced to streamline local governments and make them more effective managers of public resources by enhancing administrative efficiency and (paradoxically) placing previously public resources in private hands on the assumption that private actors would manage them more efficiently. This is an example of trying to cultivate good government at the local level. However, governance is a less restrictive concept than government, highlighting as it does the interconnectedness between the public and private spheres; the politics and processes by which citizens become involved in decision-making; as well as the sound administration of public resources by agents of government on behalf of citizens (Stoker 1998: 38). In other words, 'good governance' is not merely about the efficient allocation of public resources, but also the way decisions about allocations are made, and the ways in which citizens can hold governments accountable for their performance.

Efforts to improve governance at the urban scale have largely revolved around 'decentralisation', or the transfer of resources and responsibilities from central to local governments. From a theoretical standpoint, decentralisation is supposed to bring government closer to the governed, thereby improving information and resource flows while contributing to democratic consolidation. Proponents suggest that local governments are inherently more accountable to their constituencies than regional or national ones (World Bank 1997), and that this enhanced accountability will improve public service delivery. As a policy strategy, decentralisation has been widely adopted: it has been estimated that by the mid-1990s 80 per cent of countries were engaged in some form of decentralisation (Crook and Manor 2000); in 1994 the Urban Management Programme, funded by the World Bank and United Nations agencies, claimed that 'of the 75 developing and transitional countries with populations greater than five million, all but 12 claim to be embarked on some form of transfer of political power to local units of government' (Dillinger 1994: vii). Has decentralisation improved urban governance?

The record is mixed, reflecting both the depth and diversity of decentralisation strategies and the importance of local context. Decentralisation is a very general term that is used to describe very different kinds of changes in local–central government relations. A widely used typology of decentralisation was developed by Rondinelli (1981) who proposed that it generally takes one of three forms: deconcentration, delegation or devolution. Deconcentration refers to the geographical dispersal of central government functions. The most basic and least extensive form of decentralisation, deconcentration is basically a shifting of the central government workload outside the national capital city, allowing a greater degree of responsiveness to regions but without any transfer of authority to lower tiers of government. Delegation refers to the transferring of responsibility for decision-making and administration of public functions to other organisations, whether they be local governments or parastatal corporations. These organisations have semi-independent authority and are not entirely controlled by the central government but are nevertheless ultimately accountable to it. Finally, devolution refers to the most extensive form of decentralisation, involving an actual transfer of authority to lower tiers of government. Rather than just delegating certain responsibilities, devolution implies a significant degree of autonomy with respect to decision-making, finance and management. In such a scenario local authorities are perceived as a separate level with clear legal-geographical boundaries within which they exercise this authority and raise revenues, performing public functions with little or no direct control exercised by central authorities.

Between this diversity of decentralisation strategies and the widely varying contexts in which they have been employed, it should come as no surprise that it is difficult to determine whether or not decentralisation has been a positive or a negative force. On balance, the empirical evidence (both qualitative and quantitative) is mixed (Faguet and Sánchez 2008). Even within a single country, the effects of decentralisation can vary. For example, in a comprehensive quantitative study of the effects of decentralisation on local service delivery in Bolivia, Faguet (2003, 2004) found an overall positive impact on local government responsiveness to local needs. However, the impacts varied significantly across regions and municipalities. Further qualitative research based on interviews with local residents, business people and government officials revealed that local power dynamics ultimately determined the relative success of decentralisation across municipalities (see Box 7.2).

Critics of decentralisation are quick to point out that there are no a priori reasons why more localised forms of governance should be more accountable than at other levels (Heller 2000; Tendler 1997). Power at the local level sometimes can be more concentrated and applied more ruthlessly than when exercised at the national level. Competing interests clustered around a smaller pool of resources can exclude weaker members of society in the scramble of 'pork-barrel politics' (Beall 2005). Decentralisation policies can also be used instrumentally by politicians at higher tiers of government to win the support of recalcitrant regional or local leaders, thereby bolstering central government power in peripheral areas where opposition to a ruling party prevails. By accommodating local or regional demands for power-sharing through decentralisation strategies political opposition can be neutralised. In countries where heterogeneity spills over into regional or ethnic conflict, decentralisation can even lead to increased fragmentation and fuel hostilities (Schou and Haug 2005). Finally, for significant devolution to take place local governments need to be able to generate significant resources on their own account. Very few cities, let alone smaller district authorities, can deliver on this: 'most cities of the South have a motley collection of small local taxes which generate little revenue, cost a lot to collect, are economically distorting and impinge heavily on the poor' (Devas 1999: 11).

It might be expected, therefore, that larger cities would be better served by decentralisation processes than smaller ones. However, some recent research has arrived at the opposite conclusion. A comparative study of Bolivia and Colombia concluded that decentralisation did successfully

Box 7.2

Decentralisation in Bolivia

In 1994, Bolivia initiated major reforms designed to devolve resources and decision-making authority to lower levels of government. Overall, the programme has been assessed as a relative success, resulting in a more equitable distribution of public funds across space, more investment in poorer districts and greater responsiveness to local needs. However, results have varied significantly across districts and municipalities. An in-depth look at two districts – Viacha and Charagua – reveals the important role of local political and economic dynamics in explaining this variability.

A one-hour drive from La Paz, the administrative capital of Bolivia, the town (and district) of Viacha is almost a suburb of the city. It is, by Bolivian standards, a relatively wealthy industrial town – home to Bolivia's largest cement factory and a bottling plant owned by Bolivia's largest brewery. Together, these two firms contribute a significant proportion of the municipality's budget, and the brewery makes generous contributions in-kind through the lending of heavy machinery and donations of beer, to be used at the discretion of the mayor. However, despite the availability of a strong local revenue base and corporate largess, Viacha did not fare well under the early days of decentralisation. Corruption, wasteful public expenditures, and eventually civil unrest were regular features of the first three years under the new dispensation. These were a consequence of the political manoeuvrings of local industrial interests (particularly the brewery) and national political parties. The whole of Viacha's local government – the mayor's office, the municipal council and a citizens' oversight committee – was more or less bought by these interests and used for proxy political battles between national political parties with little interest in the welfare of the local population.

Charagua is an overwhelmingly rural district in the south-east of the country with a local economy dominated by peasant agriculture and cattle-ranching. The town of Charagua itself has essentially no industry and serves mainly as the seat of power for the local cattle-ranching elite. By 1997, Chargua's reputation was diametrically opposed to that of Viacha: a well-run district with a capable and enthusiastic mayor and a record of efficient public expenditure. These positive results stem from the local political dynamics of the district. The cattle-ranching elite class based in the town of Charagua are generally identified as white or mestizo, in contrast to the majority of the district's rural inhabitants who identify themselves as indigenous *Guaraní*. Historically, the *Guaraní* were economically and politically marginalised following their defeat at the hands of the Bolivian state's army in the late nineteenth century. But a rebirth of *Guaraní* pride and unity emerged in the 1980s in the form of the *Asamblea del Pueblo Guaraní* (APG), which established itself as an organisational mechanism for cultivating cooperation in the *Guaraní* community and representing the community's interests, both locally and internationally. This coincided with a decline in the economic might and confidence of the cattle-ranching class, who had seen their assets falling in value due to low land and commodity prices and an exodus of their well-educated children to the cities in search of better opportunities. Recognising the potential of the APG to mobilise a block of voters previously ignored, the centre-left Movimiento Bolivia Libre (MBL) party saw an opportunity to build the electoral base needed to kick out the ruling Movimiento Nacionalista Revolucionario (MNR) party, whose mayor was widely perceived as racist against the rural *Guaraní*. The MBL allowed the APG to choose its candidates for local elections in 1995 and was rewarded with a big rural turnout and electoral victory. Although still powerful, the cattle-ranching elite was not intent on controlling local politics, and the result was a mayor, municipal council and oversight committee that worked in concert to serve the interests of the poor rural majority.

Sources: Faguet (2003, 2004)

shift public investment from industry and infrastructure to social services such as health, education, water supply and sanitation, acting as a vehicle for targeting previously neglected areas and increasing government responsiveness to citizen demand. Significantly, however, the authors demonstrated that this result was driven by the smaller, poorer, rural municipalities in the sample. It appears that the process empowered 'the smallest, poorest districts disproportionately' by relocating resources from government officials at the centre to those on the periphery (Faguet and Sánchez 2008: 1311). This is one of the few studies that has sought to examine the effectiveness of decentralisation at different spatial scales, with most analyses deriving from studies of districts and small towns rather than cities.

The authors of this particular study did not attempt to explain why larger cities were less well served by decentralisation than smaller ones, but the answer may lie in the complexity of governing large, often sprawling urban agglomerations and the diversity of institutions and powerful interests at play. While some large cities are governed by a central authority, others are governed by a constellation of municipalities each concerned only with their patch of the city, making coordination difficult. Furthermore, large settlements tend to be significant sources of economic growth, raising the stakes of local decision-making processes and potentially undermining the democratic vision of decentralisation advocates. Given the demographic trends highlighted in Chapters 1 and 2, finding solutions to the challenge of governing large, rapidly expanding urban settlements will require the development of systems of metropolitan governance that effectively mediate between the needs and interests of diverse actors at different scales. Unfortunately,

> the record of metropolitan institution-building is not promising. Instead of purposeful actions creating centripetal metropolitan forces to reinforce collective problem-solving, the opposite seems to be taking place. Cumulative processes of decentralization, fragmentation, differentiation and eventual social and economic polarization in cities . . . seem to be taking place.
>
> (Cohen 2001: 55)

Moreover, the uncritical promotion of decentralisation is often at odds with the needs of large metropolitan centres. On the one hand, bringing 'government closer to the people' may sometimes strengthen relationships between government agents and members of civil society, thereby enhancing local transparency and accountability; on the other, it may result in a fragmented institutional governance structure, which impedes needed cooperation and collaboration at a larger scale thereby frustrating the 'metropolitan impulse' (Cohen 2001: 37). The metropolitan challenge is one of establishing institutional structures and processes that effectively mediate diverse interests and facilitate coordination between various levels and agencies of government to ensure comprehensive action at the metropolitan scale. South Africa, for example, has actively sought to address this challenge through the institutionalisation of integrated development plans (IDPs), which are the vehicle through which cities consult their residents about priorities and needs and align their strategic plans with those of neighbouring municipalities and government agencies at different levels (see Box 7.3).

Efforts to improve urban governance must therefore strike a balance between the potential of decentralisation and the need for consolidation. There is, clearly, no one-size-fits-all solution. Furthermore, it is important not to become too enchanted by the 'good governance' agenda, which generally promotes participatory processes and seeks to build the capacity of civil society organisations and government agencies to engage constructively. While these are perfectly sound objectives, there is a danger that participatory governance serves to depoliticise development by advancing the ideal of consensus over and above political contestation (Pieterse 2008). Formal political actors – political parties, trade unionists and conventional activists – whose commitment opened the space for local level participatory public action in the first place also can be sidelined (Harriss *et al.* 2004). The prospects for improving governance are best where there is a high degree of central state capacity, a well-developed civil society and, crucially, a political project championed by an organised political force (Heller 2001). As has been sagely observed,

Box 7.3

South Africa's integrated development plans

Following the end of apartheid and the introduction of democracy in 1994, South Africa sought to consolidate the governing of cities by establishing new metropolitan authorities in the country's major urban centres, which had previously been characterised by multiple uncoordinated municipal government structures. Alongside the creation of these new metropolitan entities a legacy of racially fragmented urban planning and governance had to be overcome. This was done through the introduction of the integrated development plan (IDP). This is the chief local planning instrument in South Africa, which espouses 'developmental local government'. It emerged in reaction to the negative effects of fragmented and project-based planning and allows municipalities to align their budgets and the implementation of projects with the strategic priorities of the national government. IDPs feed into a broader system of intergovernmental planning that aims not for 'tiers' but 'spheres' of government, where national, provincial and local government exercise particular functions and expertise within a networked governance paradigm. This system is protected by the South African Constitution (Act 108 of 1996), and in November 1996 IDPs became a legal requirement for local councils. They have to adopt a participatory approach and integrate economic, sectoral, spatial, social, institutional, environmental and fiscal strategies. The aim is 'to support the optimal allocation of scarce resources between sectors and geographical areas and across the population in a manner that provides sustainable growth, equity and the empowerment of the poor and the marginalised' (Department of Provincial and Local Government (DPLG) 2000: 15). As such, a further objective of the IDP is to assist municipalities meet their developmental mandates.

The IDP is 'the product of a peculiarly South African story' but has been shaped by various influences including the preoccupation with 'integration' in modern planning and the broadening out of town planning remits ushered in with the idea of 'urban management' in the context of the NPM (see Chapter 5). This has led critics to label the IDP process as technocratic and depoliticising, although champions see it as an important vehicle for advancing representative democracy at the local level. While the IDP is a more limited instrument of local democracy than the Brazilian innovations with participatory budgeting, the true value of the IDP has been twofold. First, there have been significant improvements in service delivery, particularly in poor areas, albeit with a degree of unevenness across the country. Second, the focus on integration promotes 'joined-up' governmental action across spheres of government and represents an advance over the emphasis on decentralisation that has characterised the international development agenda over recent decades and which is concerned primarily with shifting responsibility to the local level rather than intergovernmental coordination.

Sources: Binns and Nel (2002); DPLG (2000); Harrison (2008); Mabin (2002); Parnell *et al.* (2002); Schmidt (2008)

development is ultimately driven not only by 'good governance' but also by 'good politics' (White 1995).

Movement and inertia in urban politics

In Chapter 1, we noted that development is fundamentally about change. While effective urban planning and better governance imply normative processes and systems for effecting change, we need to know what shapes these processes and systems. In other words, where urban planning is poor or absent, or governance is characterised by exclusion, what prompts things to change for the better? The answer, of course, is politics. For example, Brazil's adoption and generally successful experience of participatory budgeting must be understood within the broader context of democratic political reform in Brazil, not simply the inherent merits of the approach itself. Indeed, although the rubric of participation has been widely adopted in the discourse of international development (as noted in Chapter 1), it is relatively rare in practice. The Brazilian experiment was born out of a generation of political contestation and struggle by marginalised groups of citizens and a social movement that ultimately transformed into a political party (see Plate 9). It is through this kind of sustained, concerted public action and political engagement that change is best catalysed and institutionalised. In Chapter 2 we suggested that urbanism is uniquely conducive to political engagement and the cultivation of citizenship due to the size, density and heterogeneity of urban populations. How does this potential manifest? In cities characterised by poverty and gross inequality, why is there not more political activity and contestation?

Studies of urban politics are divided between those who focus on urban social movements and see urban politics as defined by grassroots 'activism', and those who are preoccupied with patronage and who see urban politics as characterised by inertia and 'clientelism' (Walton 1998: 460). In reality, cities experience episodes of movement and activism on the one hand and long periods of political passivity and quiescence on the other. Characterising urban dwellers either as heroes of resistance or as inert dupes hopelessly dependent on patronage and favour is ultimately unhelpful. The interesting issue is whether 'certain conditions are conducive to more or less participation, different forms of conflict and cooperation, changing arrangements of power and so forth' (Walton 1998: 462).

Plate 9 Police confront members of an urban social movement in São Paulo

Source: Lucy Earle

In cities across the world, there are countless reasons for people to organise and express their grievances. John Walton (1998) has identified three types of collective action in response to grievances that have characterised different periods of recent history. The first is *labour action*, common in the early development decades, whereby workers engage in strikes, demonstrations and other forms of protest against unemployment and policies with adverse implications for jobs and wages. The second is *collective consumption action*, which emerges in times of high unemployment and where low-income people seek to reduce the cost of consumption of public goods such as urban services and transportation, rather than pursue futile efforts to secure wage increases. Such activities have been common since the 1980s. More recently they have been accompanied by *political and human rights action*, involving mobilisation around issues of social justice, freedom and representation, 'expressed in marches, demonstrations, vigils, hunger strikes and similar acts of conscience' (Walton 1998: 463).

Participation in social movements – loose affiliations of organisations, networks and activist individuals that connect people on the basis of shared identities or common grievances – is one of the most significant ways in which urban citizens organise around these kinds of issues. While urban social movements are inherently political in their efforts to challenge existing practices and power structures, they often seek to retain their autonomy from prevailing political agendas and to avoid alignment or affiliation with established political parties, allowing them to 'negotiate and work with whoever is in power locally, regionally or nationally' (D'Cruz and Satterthwaite 2005: 54–55). The decision to work outside conventional politics is usually highly strategic, not least because such federations and movements emerge in the first place because prevailing political arrangements do not provide legitimate avenues for their needs to be addressed (D'Cruz and Satterthwaite 2005: 57). Writing in the 1980s, the well-known theorist of urban social movements Manuel Castells recognised that some urban social movements hold on to illegality as a means of securing autonomy and avoiding co-option by the state. He noted in respect of squatters' movements in Monterrey, Mexico, that they were 'strongly opposed to the legalisation by the government of their illegal land occupation . . . as the movement continues to be illegal, it can act as a pressure group vis-à-vis the state' (Castells 1982: 275). Similarly, operating on the fringes of legality in contemporary Brazil, urban social movements such as the Movimento Sem Teto in São Paulo organise illegally to meet immediate needs as well as pursue wider political agendas (see Box 7.4).

Indeed, while urban social movements often grow out of specific local level demands for housing, services and amenities, their actions can have broader political implications. For example, some have suggested that social movements played a critical role in the re-democratisation of Brazil in the 1990s (Holston 2008; Holston and Appadurai 1993). Others,

Box 7.4

The Movimento Sem Teto in São Paulo

The Movimento Sem Teto (or Roofless Movement) is a generic term that refers to social movements campaigning for low-income housing in São Paulo city. It was inspired by the Movimento Sem Terra, or Rural Landless Workers' Movement, that has been at the forefront of campaigning for land reform since the early 1980s. Their daring occupations of unproductive ranches to set up cooperative farms has inspired the Sem Teto's strategy of urban occupations of abandoned buildings in central São Paulo.

This is one of the two main strategies employed by the housing movement, the other being to campaign for housing policy reform through input into consultative and deliberative policy forums at municipal, state and federal level. Although experiments in participatory democracy in Brazilian cities have generated much discussion amongst policy makers and academics, it is the movement's inability to achieve its aims through institutionalised participation alone that has reinforced the need for a parallel strategy on the fringes of legality – that of urban building occupations.

Through its occupations, the movement attempts to draw attention to the chronic spatial segregation of the city, which forces the poor to live in underserviced *favelas* and irregular settlements in the peripheries far from their places of work, while an estimated 40,000 buildings remain empty in the centre. This focus on the central districts presents unique challenges for the movement, given that it is a complex environment with multiple stakeholders. It has symbolic value for governments seeking to implement visible 'showcase' infrastructure projects as well as economic value for property developers. However, many buildings are dilapidated and unattractive to both businesses and higher income groups; hence they are left unoccupied.

In 2001 a piece of legislation known as the Statute of the City was passed, in which the 'right to the city' is conceived as a collective right to urban services. The Statute operationalises articles of the progressive 1988 Constitution that enshrine the right to housing. It also reinforces constitutional legislation on the social function of property, and allows for expropriation of abandoned buildings for renovation as low-income housing. The Statute and the Constitution have thus become key weapons used by the Sem Teto to support its claims, while recourse to illegal or semi-legal activity continue to be used when government legislation has not been adequately followed through. Illegal occupations therefore continue to be conceived of as a necessary corollary to engaging in formal institutions of democratic participation.

Source: Earle (2009)

however, are more cautious, questioning the ability of actually existing urban social movements to rise above local level demands to engage in scaled-up activity in sustained pursuit of broader political issues (Castells 1983: 329). In this view social movements can and do play an important role in urban public action, act as ballast to elitist tendencies in formal politics and serve to revitalise urban politics. But to do so, some argue that urban social movements need to link into an organised political force if their activities are to have wide and sustained political impact (Heller 2001).

There is also an increasing tendency for social movements to emerge in hybrid forms of collective action that defy a clean typology (such as the one presented in Box 7.4), examples of which can be found across Africa, Asia and Latin America. One of the most well known examples is the Self-Employed Women's Association (SEWA), a membership based trade union in India that was first registered in Ahmedabad in 1972 as an organisation of poor, self-employed women workers. SEWA sees itself as an organisation at the confluence of the labour movement, the cooperative movement and the women's movement. Its goal is to organise women workers, such as home-based workers and street traders, for security, employment and self-reliance. Taken up as an example of good practice by the International Labour Organization through global dissemination of the example it sets, SEWA has spawned similar organisations elsewhere and conventional trade unions are trying more seriously to organise in the informal economy. People working in the informal economy have always utilised informal social networks to enter and develop their trade, typified for example by the *Jua Kali* Associations in Nairobi (*jua kali* literally means 'under the sun' and is the name given to the informal economy in Kenya) (Mitullah 1991). However, there are burgeoning examples of larger scale organisation among urban informal workers, some more akin to global social movements than local associations (Lourenco-Lindell 2007).

For example, StreetNet International is an alliance of membership-based organisations (unions, cooperatives or associations) directly involved in organising market vendors, street traders and hawkers from all over the world, including Bangladesh, India, Nepal, Philippines, South Korea, Sri Lanka and Thailand in Asia; Bolivia, Colombia, Mexico and Peru in Latin America; and Benin, Côte d'Ivoire, Ghana, Kenya, Nigeria, Senegal, South Africa, Uganda and Zambia in Sub-Saharan Africa. In 1995 StreetNet drew up an International Declaration of Street Vendors, including a commitment to create national policies to promote and protect their rights. A longer-term objective is to revise and update the declaration to produce a new code that can be used with municipalities in cities across the world in the development of new laws and regulatory frameworks that will prevent informal traders from being the targets of state repression (Horn 2000). StreetNet is an example of scaled-up political action that remains aloof from party politics but not from political engagement and encounters with government.

There are also promising examples of local authorities engaging positively and creatively with the informal economy as a result of StreetNet affiliates. In Durban, it was estimated that the annual turnover of informal street trade vendors around the city's main public transport node was over 1 billion South African Rand, far in excess of the annual turnover of the city's largest shopping mall (Skinner 2007: 5). This helped convince the Metropolitan government of the need to shift from repression of the informal economy to greater levels of tolerance. It gave rise to a city taskforce that drew up a policy that was 'effective and inclusive' and which became operative in October 2000. In the event implementation has been somewhat selective and there has been a degree of backsliding on the part of the metropolitan government (Lund and Skinner 2004; Skinner 2007). However, this serves only to remind us that collective action, particularly on the part of more vulnerable urban dwellers, requires political allies, skilled mediators and sustained strategies.

In many ways, explaining the emergence of social movements is easier than explaining why they are not more widespread. Given the extent of urban poverty and inequality, one must ask why marginalised and excluded populations don't collectively organise more often. A number of authors have tried to explain quiescence among people who suffer conditions of incredible hardship and misery in cities (Gilbert 1994; Scheper-Hughes 1992; Wood 2003). For many low-income urban dwellers collective action is confined to concerted activities in pursuit of making ends meet, such as pooling resources, bulk buying of staples, savings clubs, income-generating activities and other self-help initiatives – activities that provide important buffers against the kind of vulnerability and risk we describe in Chapter 4. Without basic resources and security in place, it is difficult if not impossible for poor urban households to sustain self-help and mutual assistance, let alone scaled-up public action. Sometimes collective action around social needs can lead to more politicised forms of organisation but this is not automatically the case, is invariably localised and often easily suppressed. Without the safety in numbers provided by well-organised social movements (such as the Movimento Sem Teto) or enlightened urban governance (as in Durban), poor and excluded people find it difficult to claim their rights or access decision-making arenas and they are relatively powerless in formal political encounters. Not surprisingly, then, collective action at scale can be short-lived and fluid. Ephemeral groups emerge, disappear and re-form themselves but rarely remain decisively above the parapet (Beall 2001).

Sometimes what appears to be apathy can in fact be understood as loosely orchestrated resistance, or the exercise of 'the ordinary weapons of relatively powerless groups: foot dragging, dissimulation, desertion, false compliance, pilfering, feigned ignorance, slander, arson, sabotage, and so on' (J. Scott 1985: xvi). These very simple and everyday practices, often enacted by individuals, can result in incremental, almost imperceptible changes in the political environment and lay the foundation for collective action or resistance (Bayat 1997: 8). Such forms of resistance need to be recognised when seeking to explain apparent quiescence but they do not always add up to the kind of proactive local politics needed to respond effectively to the challenges of urban governance in the first urban century. Clientelism thwarts the potential for urban collection action and the vibrant urban politics necessary to generate progressive social change in cities and beyond.

If fear and powerlessness constitute one explanation for inertia in urban politics, alongside the fact that the disempowered are too busy making ends meet to engage in protest, another explanation is that the primary nature of poor people's engagement with urban politics is often via patronage. Joan Nelson's (1979) magisterial survey of urban politics in cities of Africa, Asia and Latin America concludes that rather than organised associations and radical social movements, the nature of poor people's engagement with political parties and processes is characterised by clientelism. This can take many forms. A study of poverty and urban governance conducted across ten cities in Africa, Asia and Latin America revealed examples of officials receiving private payments for the delivery of public services, and community leaders benefiting personally from such relationships through being given public sector jobs, access to government grants and a range of financial incentives in return for managing votes at election time (Devas *et al.* 2004). This practice, known as 'vote-banking', is widespread in the cities of low- and middle-income countries. In the worst cases, urban communities are intentionally denied formal tenure status so that politicians can use their insecurity as a political tool. By guaranteeing informal security through the use of political connections, a politician can rely on a block of votes from the community. If the community were regularised, this strategy would no longer work. In the best cases, poor people can use the corrupt practices to which they are subject in order to secure investments in their settlements:

> While some vote banks can rightly be criticized for being simple party-based affairs centred on liquor distribution with the poor groups being bussed to party rallies, others can also be used by poor groups to secure genuine gains from candidates in local elections. Vote bank politics can facilitate access to land and services, protect poor groups from demolition by richer ones, resolve local disputes over property boundaries, and help ensure the bureaucracy is responsive to local needs.
>
> (Mitlin 2004b: 131)

When patronage prevails in a particular city it is often the case that disadvantaged urban dwellers will show preference for clientelism over conflict 'as long as it is effective and fair' within the prevailing rules of the game (Walton 1998: 477). In other words, poor people often collude with those in power when they feel that such a strategy is most likely to ensure their short-term interests, thereby reinforcing inertia.

Unfortunately 'good politics' cannot be simply 'implemented' by those seeking to catalyse positive change. But an understanding of local and global political dynamics, of who has what to gain or lose from any particular development initiative, is essential for success. Too often development initiatives seek to disseminate 'best practice' or advocate 'good governance' without accounting for the political obstacles or opportunities in a given local context. In many cases, the best solution to a problem is just unrealistic given political constraints, and a second-best or least-worst approach is necessary. Furthermore, many important development actors, such as the UNDP, World Bank or international NGOs have neither the mandate nor desire to engage in the messy world of politics. Development interventions by such actors are always limited in their potential to generate change:

> it seems clear that the most important transformations, the changes that really matter, are not simply 'introduced' by benevolent technocrats, but fought for and made through a complex process that involves not only states and their agents, but all those with something at stake, all the diverse categories of people who craft their everyday

> tactics of coping with, adapting to, and, in various ways, resisting the established social order.
>
> (Ferguson 1994: 281)

Despite widespread poverty and inequality, and pervasive political inertia, cities ultimately offer hope for positive transformation. As noted in Chapter 2, there appears to be a broad correlation between urbanisation and democratisation. In the short run this relationship may seem spurious, but change often comes incrementally. The activities of social movements such as Movimento Sem Teto and Shack/Slum Dwellers International (see Box 7.5), as well as individuals and communities, may, over time, serve to broaden the space for political participation and representation on the part of the marginalised. Through collective and individual acts of resistance, by demanding services or the opportunity to be part of planning processes, urban residents create their own terms of citizenship firmly rooted in the fabric of the city (Holston 2008). In doing so, they help to actively shape their own political and material destinies.

Box 7.5

SDI and the Alliance in Mumbai

Established in 1996, Shack/Slum Dwellers International (SDI) is a global network of locally based organisations representing the interests of slum dwellers. Members of SDI are generally local savings groups that gradually expand their activities to include surveying their neighbourhoods, demarcating their plots, and conducting local censuses. They use these kinds of activities to build up a knowledge base that allows them to negotiate more effectively with local officials, and they utilise the global SDI network to share knowledge, raise their profile and hence improve their political bargaining potential.

The potential of SDI and its local affiliates to generate significant change is well illustrated by the activities of the Alliance in Mumbai. An SDI affiliate, the Alliance comprises three organisations that represent the interests of the urban poor in India: the Society for the Promotion of Area Resource Centres (SPARC), a registered NGO; the National Slum Dwellers' Federation (NSDF), a grassroots organisation established in the 1970s to advocate on behalf of slum dwellers; and Mahila Milan, an organisation of poor women focused on women's issues. These partners are united 'in their concern with gaining secure land tenure, adequate and durable housing and access to elements of urban infrastructure' (Appadurai 2001: 25).

In 2001, the Alliance set a major precedent. Indian Railways (IR) had undertaken a project to improve rail services, and clearing track-side slums was seen as central in improving the frequency of trains by 35 per cent. With no consultation, IR began a massive clearance campaign, sparking major demonstrations against the legality of the demolitions. This challenge was upheld by the state of Maharashtra, which had previously passed a law pronouncing that all slums built before 1995 should be deemed quasi-legal. Trying to find a suitable resolution, the state located land and approached NSDF to implement the resettlements. In the event, the NSDF peacefully and successfully relocated 60,000 railway slum dwellers.

The Alliance has been successful in many other areas as well, from designing and building public toilet blocks to securing land. Their success can be attributed in part to their political pragmatism. They avoid vote-bank politics – pervasive in Mumbai – and work with whoever is in power at local, state or national level. Theirs is a 'politics of accommodation, negotiation and long-term pressure rather than confrontation or threats of political reprisal' (Appadurai 2001: 29). Like its allies in the global SDI network, the Alliance is an instrument of 'deep democracy, rooted in local context and able to mediate globalizing forces in ways that benefit the poor' (Appadurai 2001: 23).

Sources: Appadurai (2001); Patel *et al.* (2002)

Similarly, proactive leadership, whether at the community level, local government level or national level, can motivate significant change (Cohen 2001; Friedmann 2000). In many ways, the constructive counterpart to popular mobilisation is effective leadership. As Cohen (2001) notes:

> If political leaders can put across in words and action that the diverse members of the metropolitan community have shared interests, there are possibilities for shared futures. If not, the tendencies towards fragmentation and polarisation will grow and become reflected in physical and spatial structures. In turn, this will reinforce differences which will be beyond the power of public policy to change.
>
> (Cohen 2001: 55)

The importance of a leader defining a shared vision can be seen in the example of Bogotá. The success of the TransMilenio project there (see Chapter 5) was due in large part to consecutive mayors committed to reducing congestion and integrating the city – to building the physical infrastructure for a shared future in the city. Unfortunately, this kind of leadership is in short supply. Indeed, many observers cite a generalised lack of 'political will' as a primary reason why conditions in cities in low- and middle-income countries remain static, and in many places are getting worse (UN-Habitat 2003b). Why the lack of political will?

Beyond urban bias: cities and development in the twenty-first century

The bias against urbanisation and urban development discussed in Chapter 1 persists to this day, despite the overwhelming evidence that our collective future will be defined by how we plan, manage and govern our cities. It is still evident in academic circles, international development agencies, and, perhaps most importantly, among national governments. As the intermediary between the local and the global, national governments play a critical role in determining the course of urban development. Yet the bias against urbanisation among national governments has, if anything, grown over time (see Table 7.1). In 2007, 83 per cent of African countries, 73 per cent of Asian countries and 57 per cent of Latin American countries still had policies in place to reduce rural–urban migration (United Nations 2007). These policies are almost universally unsuccessful, and even in those few cases where they worked for some time due to oppressive political measures (i.e. China and South Africa) they have proven incompatible with economic development.

One explanation for this bias against urbanisation at the national government level lies in the potential for urban spaces to stimulate political action and resistance. Across the world, cities are often home to, or even governed by, opposition political groups. Indeed, this has been a major obstacle to the implementation of real decentralisation, and to support by international agencies for city-level development initiatives: 'For international agencies to turn to the municipal arena [is] not only a technical and institutional challenge but a political challenge' (Cohen 2001: 45). Even where governments do not see support for urban development as a political threat, they often see it as undesirable,

Table 7.1 ***Anti-urbanisation policies by region, 1976–2007***

	Per cent of countries with policies to lower migration into cities			
	1976	*1986*	*1996*	*2007*
Africa	49	48	54	78
Asia	80	63	67	71
Europe	58	50	27	38
Latin America	30	68	35	68
Oceania	0	20	0	83

Source: United Nations (2007)

choosing instead to focus energy and resources on rural development, reflecting the persistence of the urban bias thesis in the minds of policy makers.

This persistence is also evident in the international community of scholars and development professionals, and is at least partially a product of a failure to fully appreciate both the pressing nature of the challenges of urbanisation, and the opportunities that urbanisation presents when seeking to combat persistent poverty, climate change and human insecurity. What is needed is a re-evaluation of the relationship between cities and development based on further research into such basic issues as how many people live in particular towns and cities, how fast they are growing, and how people access land, housing, and basic services. We need a better understanding of how national legislation, planning regulations, trade agreements and macroeconomic policies affect the lives and livelihoods of urban dwellers. We need to identify strategies for managing social, political and economic conflicts through public policy and planning practices to ensure human security in cities. We must find ways to improve the efficiency of resource use in urban areas by taking advantage of the size and density of urban settlements to reduce our impact on the environment. The persistent neglect of urban development, both at national and international levels, has resulted in 'an urban crisis' that we cannot solve without 'a full understanding of these issues' (Tannerfeldt and Ljung 2006: 163).

This is not to suggest that we do not have enough information to act now. If we are to successfully confront the most pressing challenges facing the global community today – poverty, inequality, climate change and human security – we cannot afford to wait. With a genuine commitment to urban development at all levels, from the local to the global, we have the capacity to shape an equitable, prosperous and sustainable urban future. The first step is to accept and embrace the indisputable fact that our future will indeed be an urban one.

Summary

- **The negative effects of urbanisation in nineteenth-century Europe – such as the proliferation of slums and spread of infectious diseases – gave impetus to modernist visions of top-down urban planning. By the mid-twentieth century, modernist planning had shaped cities the world over.**
- **The excesses of this approach stimulated a counter-movement advocating**

more consultative and interactive approaches to planning. Participatory budgeting, which began in Brazil but has been replicated elsewhere, is a successful example of this.

- **Urban development strategies, however, require more than just city-level planning. Integrated efforts across all levels of government, from the local to the national, are necessary.**
- **'Governance' is a broader concept than that of 'government', and incorporates the role of civil society and the private sector in decision-making, resource allocation and delivery.**
- **Efforts to improve urban governance have revolved around decentralisation. This can take many forms, and whether or not is 'successful' depends on a range of factors including scale and the ability of local authorities to generate their own revenue.**
- **Participatory urban governance, which is often associated with decentralisation, risks depoliticising urban development by sidelining formal political actors.**
- **Urban politics is often conceptualised as either being about 'activism' in the form of social movements and collective action, or 'clientelism' and inertia. Both of these perspectives tend to oversimplify the dynamics of politics in cities.**
- **Urban social movements have played critical roles in transformation but the most successful cases have involved strategic political alliances, treading a fine line between independence and co-optation by party politics.**
- **Examples of transformative collective action are relatively rare, in part due to urban dwellers' struggles for immediate survival. Engagement with urban politics thus often takes the form of patronage, which is frequently the most effective way for people to secure their short-term interests.**
- **Cities do offer the potential for deeper transformation, but inclusive urban development will always be a *political* challenge. Hence national and local political will and committed urban leadership are essential if this potential is to be realised.**

Discussion questions

1. Discuss the strengths and weaknesses of participatory budgeting as an approach to urban planning.
2. Under what conditions is decentralisation likely to be good for urban development?
3. Are 'formal' political actors such as political parties and trade unions important for urban development? Why?

4 Discuss the potential for collective action in cities to bring about transformative political change, with reference to urban social movements.
5 What are the main obstacles to integrated strategies for urban development based on cooperation between state agencies at different scales? What role should the *national* state play in addressing urban challenges?

Further reading

Birch, Eugénie (ed.) (2008) *The Urban and Regional Planning Reader*, London: Routledge.

Castells, Manuel (1983) *The City and the Grassroots: A Cross-Cultural Theory of Urban Social Movements*, Berkeley, CA: University of California Press.

Devas, N. with P. Amis, J. Beall, U. Grant, D. Mitlin, F. Nunan and C. Rakodi (2004) *Urban Governance, Voice and Poverty in the Developing World*, London: Earthscan.

Faguet, Jean-Paul and Fabio Sánchez (2008) 'Decentralization's effects on educational outcomes in Bolivia and Colombia', *World Development*, 36(7): 1294–1316.

Friedmann, John (2007) 'The wealth of cities: towards an assets-based development of newly urbanizing regions', *Development and Change*, 38(6): 987–998.

Hall, Peter (2002) *Cities of Tomorrow*, 3rd edn, Oxford: Blackwell.

Heller, P. (2001) 'Moving the state: the politics of democratic decentralisation in Kerala, South Africa and Porto Alegre', *Politics and Society*, 29(1): 131–163.

Strom, Elizabeth and John H. Mollenkopf (2006) *The Urban Politics Reader*, London: Routledge.

Walton, John (1998) 'Urban conflict and social movements in poor countries: theory and evidence of collective action', *International Journal of Urban and Regional Research*, 22(3): 460–481.

Useful websites

Cities Alliance, a global coalition of cities and their development partners: www.citiesalliance.org/index.html

Global Development Research Center (GDRC) Programme on Urban Governance: www.gdrc.org/u-gov/ugov-define.html

Self-Employed Women's Association (SEWA): www.sewa.org

Shack/Slum Dwellers International (SDI): www.sdinet.co.za

References

Abhat, D., S. Dineen, T. Jones, J. Motavalli, R. Sanborn and K. Slomkowski (2005) 'Cities of the future: today's "mega-cities" are overcrowded and environmentally stressed', *E, the Magazine of the Environment*, October, available at www.emagazine.com/view/?2849 (accessed 4 February 2009).

Acemoglu, D., S. Johnson and J.A. Robinson (2001) 'The colonial origins of comparative development: an empirical investigation', *American Economic Review*, 91(5): 1369–1401.

Ades, Alberto F. and Edward L. Glaeser (1995) 'Trade and circuses: explaining urban giants', *Quarterly Journal of Economics*, 110(1): 195–227.

Agostini, G., F. Chianese, W. French and A. Sandhu (2007) *Understanding the Processes of Urban Violence: An Analytical Framework*, report prepared for the Cities and Conflict Theme of the Crisis States Research Centre, London School of Economics.

Alam, Mozaharul and M.D. Golam Rabbani (2007) 'Vulnerabilities and responses to climate change for Dhaka', *Environment and Urbanization*, 19(1): 81–97.

Allmendinger, Philip (2002) *Planning Theory*, London: Palgrave Macmillan.

Amsden, Alice (2001) *The Rise of 'The Rest': Challenges to the West from Late-Industrializing Economies*, Oxford: Oxford University Press.

Anderson, David and Richard Rathbone (2000) 'Urban Africa: histories in the making', in David Anderson and Richard Rathbone (eds) *Africa's Urban Past*, Oxford: James Currey and Heinemann.

Appadurai, Arjun (1996) *Modernity at Large: Cultural Dimensions of Globalization*, Minneapolis, MN: University of Minnesota Press.

Appadurai, Arjun (2001) 'Deep democracy: urban governmentality and the horizon of politics', *Environment and Urbanization*, 13(2): 23–43.

Appadurai, Arjun (2006) *Fear of Small Numbers: An Essay on the Geography of Anger*, Durham, NC: Duke University Press.

Arnott, Richard J. and Mark Gersovitz (1986) 'Social welfare underpinnings of urban bias and unemployment', *Economic Journal*, 96(382): 413–424.

Ashworth, Gregory (1991) *War and the City*, London: Routledge.

Bairoch, Paul (1988) *Cities and Economic Development: From the Dawn of History to the Present*, Chicago, IL: University of Chicago Press.

Bako, Sabo (2007) 'The rise and expansion of motor cycle business and its associational activism in Nigeria', paper presented to the Conference on Informalizing Economies and New Organizing Strategies in Africa held at the Nordic African Institute, Uppsala, 20–22 April.

Banarji, Gautam (1997) 'The impact of modern warfare: the case of Iraq', in J. Beall (ed.) *A City for All: Valuing Difference and Working with Diversity*, London: Zed.

Bannister, Jon and Nick Fyfe (2001) 'Introduction: fear and the city', *Urban Studies*, 38(5–6): 807–813.

Baran, Paul (1957) *The Political Economy of Growth*, New York: Monthly Review Press.

Bartlett, Sheridan (2003) 'Water, sanitation and urban children: the need to go beyond "improved" provision', *Environment and Urbanization*, 15(2): 56–58.

Basant, Rakesh and Pankaj Chandra (2007) 'Role of educational and R&D institutions in city clusters: an exploratory study of Bangalore and Pune regions in India', *World Development*, 35(6): 1037–1055.

Bates, Robert (1981) *Markets and States in Tropical Africa*, Berkeley, CA: University of California Press.

Bates, Robert (1988) *Toward a Political Economy of Development*, Berkeley, CA: University of California Press.

Batley, Richard (1996) 'Public–private relationships and performance in service provision', *Urban Studies*, 33(4–5): 723–751.

Batley, Richard and George Larbi (2004) *The Changing Role of Government: The Reform of Public Services in Developing Countries*, Basingstoke: Palgrave Macmillan.

Bayat, Asef (1997) *Street Politics: Poor People's Movements in Iran*, New York: Columbia University Press.

BBC News (2008) 'Riots prompt Ivory Coast tax cuts', 2 April, available at news.bbc.co.uk/2/hi/africa/7325733.stm (accessed 4 February 2009).

Beall, Jo (1995) 'Social security and social networks among the urban poor in Pakistan', *Habitat International*, 19(4): 427–455.

Beall, Jo (1997a) 'Social capital in waste: a solid investment?' *Journal of International Development*, 9(7): 951–961.

Beall, Jo (1997b) 'Thoughts on poverty from a South Asian rubbish dump: gender, inequality and waste', *IDS Bulletin*, 28(3): 73–90.

Beall, Jo (2001) 'Valuing social resources or capitalising on them? The limits to pro-poor urban governance in nine cities of the South', *International Planning Studies*, 6(4): 357–375.

Beall, Jo (2002) 'Living in the present, investing in the future – household security among the poor', in Carole Rakodi with Tony Lloyd-Jones (eds) *Urban Livelihoods: A People-Centred Approach to Reducing Poverty*, London: Earthscan.

Beall, Jo (2004) 'Surviving in the city: livelihoods and linkages of the urban poor', in N. Devas with P. Amis, J. Beall, U. Grant, D. Mitlin, F. Nunan and C. Rakodi, *Urban Governance, Voice and Poverty in the Developing World*, London: Earthscan.

Beall, Jo (2005) *Funding Local Governance: Small Grants for Democracy and Development*, London: ITDG Publishing.
Beall, Jo (2006) 'Cities, terrorism and development', *Journal of International Development*, 18(1): 105–120.
Beall, Jo (2007) *Cities, Terrorism and Urban Wars of the 21st Century*, Crisis States Programme Series Two, Working Paper no. 9, London: Crisis States Development Research Centre, Development Studies Institute, London School of Economics, February.
Beall, Jo and Daniel Esser (2005) *Shaping Urban Futures: Challenges to Governing and Managing Afghan Cities*, Kabul: Afghanistan Research and Evaluation Unit, March.
Beall, Jo and Stefan Schıtte (2006) *Urban Livelihoods in Afghanistan*, Kabul: Afghanistan Research and Evaluation Unit, August.
Beall, Jo, Owen Crankshaw and Susan Parnell (2000) 'Victims, villains and fixers: the urban environment and Johannesburg's poor', *Journal of Southern African Studies*, 26(4): 803–855.
Beall, Jo, Owen Crankshaw and Susan Parnell (2002) *Uniting a Divided City: Governance and Social Exclusion in Johannesburg*, London: Earthscan.
Beall, Jo, Thomas Goodfellow and James Putzel (2006) 'Introductory article: on the discourse of terrorism, security and development', *Journal of International Development*, 18(1): 51–68.
Beaverstock, J.V., R.G. Smith and P.J. Taylor (1999) 'A roster of world cities', *Cities*, 16(6): 445–458.
Bebbington, Anthony (1999) 'Capitals and capabilities: a framework for analyzing peasant viability, rural livelihoods and poverty', *World Development*, 27(12): 2021–2044.
Becker, C.M. and A.R. Morrison (1999) 'Urbanization in transforming economies', in P. Cheshire and E.S. Mills (eds) *Handbook of Regional and Urban Economics*, Vol. 3, *Applied Urban Economics*, Oxford: Elsevier.
Berman, Marshall (1987) 'Among the ruins', *New Internationalist*, 178 (December): 1–3.
Bernstein, Henry (2000) 'Colonialism, capitalism, development', in Tim Allen and Alan Thomas (eds) *Poverty and Development in the 21st Century*, Oxford: Oxford University Press.
Binns, Tony and Etienne Nel (2002) 'Devolving development: integrated development planning and developmental local government in post-apartheid South Africa', *Regional Studies*, 36(8): 921–945.
Bollens, Scott (2001) 'City and soul – Sarajevo, Johannesburg, Jerusalem, Nicosia', *City*, 5(2): 169–187.
Bollens, Scott (2006) 'Urban planning and peace building', *Progress in Planning*, 66: 67–139.
Bond, Patrick (2000) *Cities of Gold, Townships of Coal: Essays on South Africa's New Urban Crisis*, Trenton, NJ: Africa World Press.
Borja, Jordi and Manuel Castells (1997) *Local and Global: The Management of Cities in the Information Age*, London: Earthscan.

Boserup, Ester (1970) *Women's Role in Economic Development*, London: Earthscan.

Bosman, Martin (2005) 'A review of privatization of water and sanitation systems: the case of Greater Buenos Aires', draft discussion paper, prepared for the Conference on Global Cities: Water, Infrastructure and Environment, Santa Barbara, CA, June, available at www.globalization-africa.org/papers/56.pdf (accessed 4 February 2009).

Bradley, D., C. Stephens, T. Harpham and S. Cairncross (1992) *A Review of Environmental Health Impacts in Developing Country Cities*, Washington, DC: World Bank.

Braudel, Fernand (1984) *Civilization and Capitalism, 15th–18th Century*, Vol. 3, *The Perspectives of the World*, New York: Harper & Row.

Brennan-Galvin, Ellen (2002) 'Crime and violence in an urbanizing world', *Journal of International Affairs*, 56(1): 123–145.

Budds, Jessica and Gordon McGranahan (2003) 'Are the debates on water privatization missing the point? Experiences from Africa, Asia and Latin America', *Environment and Urbanization*, 15(2): 87–114.

Butler, Stuart (1991) 'Privatization for public purposes', in William T. Gormley Jr. (ed.) *Privatization and its Alternatives*, Madison, WI: University of Wisconsin Press.

Butterworth, Douglas and John K. Chance (1981) *Latin American Urbanization*, Cambridge: Cambridge University Press.

Cabannes, Yves (2003) 'Participatory budgeting and municipal finance: participatory perspectives in cities of Europe and Latin America', *Urban Network Seminar no. 9*, Porto Alegre, Brazil: UN-Habitat.

Caldeira, Theresa (2000) *City of Walls: Crime, Segregation and Citizenship in São Paolo*, Berkeley, CA: University of California Press.

Caplan, Richard (2005) *International Governance of War-Torn Territories: Rule and Reconstruction*, Oxford: Oxford University Press.

Cardoso, Fernando H. (1972) 'Dependency and development in Latin America', *New Left Review*, 74: 83–95.

Carney, Diana (1998) 'Implementing the sustainable livelihoods approach', in Diana Carney (ed.) *Sustainable Rural Livelihoods: What Contribution Can We Make?* London: Department for International Development.

Castells, Manuel (1982) 'Squatters and politics in Latin America: a comparative analysis of urban social movements in Chile, Peru and Mexico', in Helen I. Safa (ed.) *Towards a Political Economy of Urbanization in Third World Countries*, Delhi: Oxford University Press.

Castells, Manuel (1983) *The City and the Grassroots: A Cross-Cultural Theory of Urban Social Movements*, Berkeley, CA: University of California Press.

Castells, Manuel (2000) 'Urban sociology in the twenty-first century', in Ida Susser (ed.) *The Castells Reader on Cities and Social Theory*, Oxford: Blackwell.

Castells, Manuel (2002) 'Preface: sustainable cities: structure and agency', in P. Evans (ed.) *Livable Cities? Urban Struggles for Livelihood and Sustainability*, Berkeley, CA: University of California Press.

Castells, Manuel and Alejandro Portes (1989) 'World underneath: the origins, dynamics and effects of the informal economy' in Alejandro Portes, Manuel Castells and Lauren A. Benton (eds) *The Informal Economy: Studies in Advanced and Less Developed Countries*, Baltimore, MD: Johns Hopkins University Press.

Castro, José Esteban (2005) 'Water-borne diseases', in Tim Forsyth (ed.) *Encyclopedia of International Development*, London: Routledge.

Chambers, Robert (1983) *Rural Development: Putting the Last First*, Harlow, UK: Longmans.

Chambers, Robert and Gordon Conway (1992) *Sustainable Rural Livelihoods: Practical Concepts for the 21st Century*, IDS Discussion Paper no. 296, Brighton: Institute of Development Studies, University of Sussex.

Chandhoke, N., P. Priyadarshi, S. Tyagi and N. Khanna (2007) 'The displaced on Ahmedabad', *Political and Economic Weekly*, 27 October: 10–14.

Chang, Ha-Joon (2003) *Globalisation, Economic Development and the Role of the State*, London: Zed.

Childe, V. Gordon (1950) 'The Urban Revolution', *Town Planning Review*, 21(1): 3–17.

Choguill, Charles (1995) 'The future of planned urban development in the Third World: new directions', in Brian Aldrich and Ranvinder Sandhu (eds) *Housing the Urban Poor: Policy and Practice in Developing Countries*, London: Zed.

Cochrane, Allan (2007) *Understanding Urban Policy: A Critical Approach*, Oxford: Blackwell.

Cohen, Michael (2001) 'Urban assistance and the material world: learning by doing at the World Bank', *Environment and Urbanization*, 13(1): 37–60.

COHRE (2006) *Global Survey on Forced Evictions: Violations of Human Rights*, Geneva: Centre on Housing Rights and Evictions, December, available at www.cohre.org/store/attachments/Global_Survey_10.pdf (accessed 4 February 2009).

Collier, Paul (2000) 'Doing well out of war: an economic perspective', in Mats Berdal and David Malone (eds) *Greed and Grievance: Economic Agendas in Civil Wars*, Boulder, CO: Lynne Rienner.

ComHabitat (2005) *PRSPs, Human Settlements and Urban Poverty*, Commonwealth Action for Human Settlements, *ComHabitat Brief 1*, London: Commonwealth Secretariat.

Cooper, Fredrick (2002) *Africa since 1940: The Past of the Present*, New York: Cambridge University Press.

Corbridge, Stuart and Gareth Jones (2006) *The Continuing Debate about Urban Bias: The Thesis, its Critics, its Influence, and Implications for Poverty Reduction*, London: Department of Geography and Environment, London School of Economics and Political Science.

Cornia, Giovanni Andrea, Richard Jolly and Frances Stewart (eds) (1987) *Adjustment with a Human Face*, Volume 1, Oxford: Oxford University Press.

Coward, Martin (2004) 'Urbicide in Bosnia', in Stephen Graham (ed.) *Cities, War and Terrorism: Towards an Urban Geopolitics*, Oxford: Blackwell.

Cowen, Michael and Robert Shenton (1996) *Doctrines of Development*, London: Routledge.

Crankshaw, Owen, Alan Gilbert and Alan Morris (2000) 'Backyard Soweto', *International Journal of Urban and Regional Research*, 24(4): 841–857.

Crook, Richard and James Manor (2000) *Democratic Decentralization*, Operations Evaluation Department Working Paper, Washington, DC: World Bank.

Davis, Diane (2007) 'What kind of conflict? Cities, war and the failure of urban public security', in Human Security – Cities, *Human Security for an Urban Century: Local Challenges, Global Perspectives*, Ottawa: Foreign Affairs and International Trade Canada, March.

Davis, J.C. and J.V. Henderson (2003) 'Evidence on the political economy of the urbanization process', *Journal of Urban Economics*, 53: 98–125.

Davis, Mike (1990) *City of Quartz: Excavating the Future in Los Angeles*, London: Verso.

Davis, Mike (1998) *Ecology of Fear, Los Angeles and the Imagination of Disaster*, London: Picador.

Davis, Mike (2006) *Planet of Slums*, London: Verso.

D'Cruz, Celine and David Satterthwaite (2005) *Building homes, Changing Official Approaches: The Work of Urban Poor Organizations and their Federations and their Contributions to Meeting the Millennium Development Goals in Urban Areas.* Poverty Reduction in Urban Areas Series, Working Paper 16, London: International Institute for Environment and Development.

Department of Provincial and Local Government (DPLG) (2000) *A Policy Paper on Integrated Development Planning*, Pretoria: DPLG.

de Soto, Hernando (1989) *The Other Path: The Economic Answer to Terrorism*, New York: Basic Books.

de Soto, Hernando (2001) *The Mystery of Capital*, London: Black Swan.

Devas, Nick (1999) *Who Runs Cities? The Relationship between Urban Governance, Service Delivery and Poverty*, Urban Governance, Partnership and Poverty Working Paper 4, Birmingham: University of Birmingham.

Devas, N. with P. Amis, J. Beall, U. Grant, D. Mitlin, F. Nunan and C. Rakodi (2004) *Urban Governance, Voice and Poverty in the Developing World*, London: Earthscan.

De Waard, Jaap (1999) 'The private security industry in international perspective', *European Journal on Criminal Policy and Research*, 7: 143–174.

Dillinger, William (1994) *Decentralization and its Implications for Urban Service Delivery*, Urban Management Programme, New York: UNDP/UNCHS/World Bank.

Donkor, Kwabena (2002) 'Structural adjustment and mass poverty in Ghana', in Peter Townsend and David Gordon (eds) *World Poverty, New Policies to Defeat an Old Enemy*, Bristol: Policy Press.

Dowdney, Luke (2004) *Neither War Nor Peace: International Comparisons of Children and Youth in Organised Armed Violence*, Rio de Janeiro: Viva Rio and Instituto de Estudos da Religião.

Drakakis-Smith, David (2000) *Third World Cities*, 2nd edn, London: Routledge.
Duranton, Gilles and Diego Puga (2001) 'Nursery cities: urban diversity, process innovation, and the life cycle of products', *American Economic Review*, 91(5): 1454–1477.
Dyson, Tim (2001) 'A partial theory of world development: the neglected role of the demographic transition in the shaping of modern society', *International Journal of Population Geography*, 7: 1–24.
Earle, Lucy (2009) 'Occupying the illegal city: urban social movements and social citizenship in Sao Paulo', PhD thesis, London School of Economics and Political Science.
Elson, Diane (ed.) (1991) *Male Bias in the Development Process*, Manchester: Manchester University Press.
Engels, Friedrich (1845) *The Conditions of the Working Class in England*, Marx/Engels Internet archive, available at www.marxists.org/archive/marx/works/1845/condition-working-class/index.htm (accessed 4 February 2009).
Escobar, Arturo (1995) *Encountering Development: The Making and Unmaking of the Third World*, Princeton, NJ: Princeton University Press.
Esser, Daniel (2004) 'The city as arena, hub and prey – patterns of violence in Kabul and Karachi', *Environment and Urbanization*, 16(2): 31–38.
Esser, Daniel (2007) 'Target Kabul: human insecurity in the Afghan capital', in Human Security – Cities, *Human Security for an Urban Century: Local Challenges, Global Perspectives*, Ottawa: Foreign Affairs and International Trade Canada.
Estache, Antonio and Martin Rossi (2002) 'How different is the efficiency of public and private water companies in Asia?', *World Bank Economic Review*, 16(1): 139–148.
Evans, Peter (1995) *Embedded Autonomy: States and Industrial Transformation*, Princeton, NJ: Princeton University Press.
Eviatar, Daphne (2004) 'Africa's oil tycoons', *The Nation*, 12 April.
Faguet, Jean-Paul (2003) *Decentralization and Local Government in Bolivia: An Overview from the Bottom Up*, Crisis States Programme Series One, Working Paper no. 29, London: Crisis States Research Centre, Development Studies Institute, London School of Economics.
Faguet, Jean-Paul (2004) 'Does decentralization increase government responsiveness to local needs? Evidence from Bolivia', *Journal of Public Economics*, 88(3–4): 867–893.
Faguet, Jean-Paul and Fabio Sánchez (2008) Decentralization's effects on educational outcomes in Bolivia and Colombia', *World Development*, 36(7): 1294–1316.
Fajnzylber, Pablo, Daniel Lederman and Norman Loayza (2002) 'Inequality and violent crime', *Journal of Law and Economics*, 45(1): 1–40.
Fay, Marianne and Charlotte Opal (2000) *Urbanization without Growth: A Not-So-Uncommon Phenomenon*, World Bank Policy Research Working Paper 2412, Washington, DC: World Bank.
Fazal, Shahab (2006) 'Addressing congestion and transport-related air pollution in Saharanpur, India', *Environment and Urbanization*, 18(1): 141–154.

Feige, Edgar L. (1990) 'Defining and estimating underground and informal economies: the new institutional economics approach', *World Development*, 18(7): 989–1002.

Ferguson, James (1990) *The Anti-Politics Machine: 'Development', Depoliticization and Bureaucratic Power in Lesotho*, Cambridge: Cambridge University Press.

Ferguson, James (1994) *The Anti-Politics Machine: 'Development', Depoliticization, and Bureaucratic Power in Lesotho*, Minneapolis, MN: University of Minnesota Press.

Foucault, Michel (1988) *Politics, Philosophy, Culture: Interviews and Other Writings, 1977–1984*, edited by Lawrence Kritzman, New York: Routledge.

Frank, Andre Gunder (1967) *Capitalism and Underdevelopment in Latin America: Historical Studies of Chile and Brazil*, New York: Monthly Review Press.

Freund, Bill (2007) *The African City*, Cambridge: Cambridge University Press.

Friedmann, John (1967) 'Regional planning and nation-building: an agenda for international research', *Economic Development and Cultural Change*, 16(1): 119–129.

Friedmann, John (1986) 'The world city hypothesis', *Development and Change*, 17: 69–84.

Friedmann, John (2000) 'The good city: in defense of Utopian thinking', *International Journal of Urban and Regional Research*, 24(2): 460–472.

Friedmann, John (2005) *China's Urban Transition*, Minneapolis, MN: University of Minnesota Press.

Friedmann, John (2007) 'The wealth of cities: towards an assets-based development of newly urbanizing regions', *Development and Change*, 38(6): 987–998.

Friedmann, John and William Alonso (eds) (1975) *Regional Policy: Readings in Theory and Applications*, Cambridge, MA: MIT Press.

Friedmann, John and G. Wolf (1982) 'World city formation: an agenda for research and action', *International Journal of Urban and Regional Research*, 15(1): 269–283.

Friere, Mila and Mario Polèse (2003) *Connecting Cities with Macroeconomic Concerns: The Missing Link*, Washington, DC: World Bank.

Galor, Oded (2005) 'The demographic transition and the emergence of sustained economic growth', *Journal of the European Economic Association*, 3(2–3): 494–504.

Gandy, Matthew (2006) 'Planning, anti-planning and the infrastructure crisis facing metropolitan Lagos', *Urban Studies*, 43(2): 371–396.

Gilbert, Alan (1994) *The Latin American City*, Nottingham: Russell Press.

Gilbert, Alan and Owen Crankshaw (1999) 'Comparing South African and Latin American experience: migration and housing mobility', *Urban Studies*, 36(1): 2375–2400.

Gilbert, Alan and Josef Gugler (1992) *Cities, Poverty and Development: Urbanization in the Third World*, 2nd edn, Oxford: Oxford University Press.

Gilbert, Alan and Anne Varley (1990) 'Renting a home in a third World city:

choice or constraint?', *International Journal of Urban and Regional Research*, 14(1): 89–108.

Glaeser, Edward L. and Bruce Sacerdote (1999) 'Why is there more crime in cities?', Part 2: Symposium on the Economic Analysis of Social Behavior in Honor of Gary S. Becker, *Journal of Political Economy*, 107(6): S225–S258.

Glaeser, E.L., H.D. Kallal, J.A. Scheinkman and A. Schleifer (1992) 'Growth in cities', *Journal of Political Economy*, 100: 1126–1152.

Goldstein, Daniel (2004) *The Spectacular City: Violence and Performance in Urban Bolivia*, Durham, NC: Duke University Press.

Graham, Stephen (2003) 'Lessons in urbicide', *New Left Review*, 19 (January–February): 63–78.

Graham, Stephen (2004a) 'Introduction: cities, warfare, and states of emergency', in Stephen Graham (ed.) *Cities, War and Terrorism: Towards an Urban Geopolitics*, Oxford: Blackwell.

Graham, Stephen (2004b) 'Cities as strategic sites: place annihilation and urban geopolitics', in Stephen Graham (ed.) *Cities, War and Terrorism: Towards an Urban Geopolitics*, Oxford: Blackwell.

Graham, Stephen (2004c) 'Constructing urbicide by bulldozer in the Occupied Territories', in Stephen Graham (ed.) *Cities, War and Terrorism: Towards an Urban Geopolitics*, Oxford: Blackwell.

Graham, Stephen and Simon Marvin (2001) *Splintering Urbanism: Networked Infrastructures, Technological Mobilities and the Urban Condition*, London: Routledge.

Gugler, Josef (ed.) (1988) *The Urbanization of the Third World*, Oxford: Oxford University Press.

Gugler, Josef (ed.) (2004) *World Cities beyond the West: Globalization, Development and Inequality*, Cambridge: Cambridge University Press.

Haddad, Lawrence, Marie T. Ruel and James L. Garrett (1999) 'Are urban poverty and undernutrition growing? Some newly assembled evidence', *World Development*, 27(11): 1891–1904.

Hägerstrand, Torsten (1978) 'An equitable urban structure', in L.S. Bourne and J.W. Simmons (eds) *Systems of Cities: Readings on Structure, Growth and Policy*, New York: Oxford University Press.

Hall, Peter (1966) *The World Cities*, London: Weidenfeld & Nicolson.

Hall, Peter (1998) *Cities in Civilization*, London: Weidenfeld & Nicolson.

Hall, Peter (2002) *Cities of Tomorrow*, 3rd edn, Oxford: Blackwell.

Hanson, Gordon H. (2001) 'U.S.–Mexico integration and regional economies: evidence from border-city pairs,' *Journal of Urban Economics*, 50(2): 259–287.

Hardoy, Jorge and David Satterthwaite (1989) *Squatter Citizen: Life in the Urban Third World*, London: Earthscan.

Hardoy, Jorge, Diana Mitlin and David Satterthwaite (2001) *Environmental Problems in an Urbanizing World: Finding Solutions in Cities in Africa, Asia and Latin America*, London: Earthscan.

Harris, Clive (2003) *World Bank Participation in Infrastructure in Developing Countries: Trends, Impacts and Policy Lessons*, World Bank Working Paper 5, Washington, DC: World Bank.

Harris, John R. and Michael Todaro (1970) 'Migration, unemployment and development: a two-sector analysis', *American Economic Review*, 60 (March): 126–142.

Harrison, Philip (2008) 'The origins and outcomes of South Africa's integrated development plans', in M. van Donk, M. Swilling, E. Pieterse and S. Parnell (eds) *Consolidating Developmental Local GovernmentL Lessons from the South African Experience*, Cape Town: University of Cape Town Press.

Harriss, John, Kristan Stokke and Olle Törnqvist (eds) (2004) *Politicising Democracy: The New Local Politics and Democratisation*, Basingstoke: Palgrave.

Hart, Keith (1973) 'Informal income opportunities and urban employment in Ghana', *Journal of Modern African Studies*, 11(1): 61–89.

Harvey, Charles (1991) 'Recovery from macro-economic disaster in Sub-Saharan Africa', in Christopher Colclough and James Manor (eds) *States or Markets? Neo-liberalism and the Development Policy Debate*, Oxford: Clarendon Press.

Healey, Patsy (1997) *Colalborative Planning: Shaping Places in Fragmented Societies*, Vancouver: University of British Columbia Press.

Heisler, Helmuth (1971) 'The creation of a stabilized urban society: a turning point in the development of Northern Rhodesia/Zambia', *African Affairs*, 70(279): 125–145.

Heller, Patrick (2000) 'Degrees of democracy: some comparative lessons from India', *World Politics*, 52(4): 484–519.

Heller, Patrick (2001) 'Moving the state: the politics of decentralization in Kerala, South Africa and Porto Alegre', *Politics and Society*, 29(1): 131–163.

Henderson, J.V., A. Kuncoro and M. Turner (1995) 'Industrial development in cities', *Journal of Political Economy*, 103: 1067–1085.

Henderson, Vernon (2003) 'The urbanization process and economic growth: the so-what question', *Journal of Economic Growth*, 8: 47–71.

Hills, Alice (2004) 'Continuity and discontinuity: the grammar of urban military operations', in Stephen Graham (ed.) *Cities, War and Terrorism: Towards an Urban Geopolitics*, Oxford: Blackwell.

Hinderink, Jan and Milan Titus (2002) 'Small towns and regional development: major findings and policy implications from comparative research', *Urban Studies*, 39(3): 379–391.

Hirschman, Albert O. (1958) *Strategy of Economic Development*, New Haven, CT: Yale University Press.

Hirschmann, David (1993) 'Institutional development in the era of economic policy reform: concerns, contradictions and illustrations from Malawi', *Public Administration and Development*, 13(2): 113–128.

Hobsbawm, Eric (2005) 'Cities and insurrection', *Global Urban Development*, 1(1): 1–8.

Holston, J. (2008) *Insurgent Citizenship: Disjunctions of Democracy and Modernity in Brazil*, Princeton, NJ: Princeton University Press.

Holston, J. and A. Appadurai (1993) 'Cities and citizenship', in. J. Holston (ed.) *Cities and Citizenship*, Durham, NC: Duke University Press.

Hood, Christopher (1991) 'A public management for all seasons?', *Public Administration*, 69: 3–19.
Horn, Pat (2000) 'What is the StreetNet Association?', *StreetNet International Newsletter No. 1*, December, available at www.streetnet.org.za/english/newsa1.htm (accessed 4 February 2009).
Human Security – Cities (2007) *Human Security for an Urban Century: Local Challenges, Global Perspectives*, Ottawa: Foreign Affairs and International Trade Canada.
Huq, Saleemul and David Satterthwaite (2008) 'Editorial: climate change and cities', *ID21 Insights*, January, available at www.id21.org/insights/insights71/art00.html (accessed 4 February 2009).
ICWE (1992) *The Dublin Statement on Water and Sustainable Development*, Dublin: International Conference on Water and the Environment, available at www.gdrc.org/uem/water/dublin-statement.html (accessed 4 February 2009).
Illich, Ivan (1992) 'Needs', in Wolfgang Sachs (ed.) *The Development Dictionary*, London: Zed.
ILO (2002) *Women and Men in the Informal Economy: A Statistical Picture*, Geneva: International Labour Office.
ILO (2007) *Global Employment Trends Brief, January 2007*, International Labour Office, available at www.ilo.org/public/english/employment/strat/global.htm (accessed 4 February 2009).
InfoChange News (2005) 'The price the poor pay in Mumbai', *InfoChange News and Features*, October, available at: www.infochangeindia.org/agenda3_06.jsp (accessed 4 February 2009).
Jacobs, Jane (1961) *The Death and Life of Great American Cities*, London: Penguin.
Jacobs, Jane (1969) *The Economy of Cities*, New York: Random House.
Jacobs, Jane (1984) *Cities and the Wealth of Nations*, New York: Vintage.
Jellinek, Lea (2000) 'Jakarta, Indonesia, Kampung culture or consumer culture?', in N. Low, B. Gleeson, I. Elander and R. Lidskog (eds) *Consuming Cities: The Urban Environment in the Global Economy after the Rio Declaration*, London: Routledge.
Jørgensen, Steen Lau and Julie Van Domelen (2001) 'Helping the poor manage risk better: the role of social funds', in N. Lustig (ed.) *Shielding the Poor: Social Protection in the Developing World*, Washington, DC: Brookings Institution Press and Inter-American Development Bank.
Joy, Claire and Peter Hardstaff (2005) *Dirty Aid, Dirty Water: The UK Government's Push to Privatise Water and Sanitation in Poor Countries*, London: World Development Movement.
Kabeer, Naila (1994) *Reversed Realities: Gender Hierarchies in Development Thought*, London: Verso.
Kaldor, Mary (2006) *New and Old Wars: Organized Violence in a Global Era*, 2nd edn, Oxford: Polity Press.
Kanji, Nazneen (2002) 'Social funds in sub-Saharan Africa: how effective for poverty reduction?', in Peter Townsend and David Gordon (eds) *World Poverty: New Policies to Defeat an Old Enemy*, Bristol: Policy Press.

King, Anthony D. (1976) *Colonial Urban Development: Culture, Social Power and Environment*, London: Routledge & Kegan Paul.

King, Anthony D. (1990) *Urbanism, Colonialism and the World-Economy*, London: Routledge.

Kironde, Lusugga J.M. (1999) 'The governance of waste management in African cities', in A. Atkinson, J. Dávila, E. Fernandes and M. Mattingly (eds) *The Challenge of Environmental Management in Urban Areas*, Ashgate Studies in Environmental Policy and Practice, Aldershot: Ashgate.

Knox, Paul and Peter Taylor (eds) (1995) *World Cities in a World System*, Cambridge: Cambridge University Press.

Kohli, Atul (2004) *State Directed Development, Political Power and Industrialization in the Global Periphery*, Cambridge: Cambridge University Press.

Kumar, Sunil (1996) 'Landlordism in Third World urban low-income settlements: a case for further research', *Urban Studies*, 33(4–5): 753–782.

Kumar, Sunil (2001) 'Embedded tenures: private renting and housing policy in India', *Housing Studies*, 16(4): 425–442.

Kumar, Sunil (2002) 'Round pegs and square holes: mismatches between poverty and housing policy in urban India', in Peter Townsend and David Gordon (eds) *World Poverty: New Policies to Defeat an Old Enemy*, Bristol: Policy Press.

Lal, Deepak (1985) *The Poverty of 'Development Economics'*, Cambridge, MA: Harvard University Press.

Lall, Somik V., Harris Selod and Zmarak Shalizi (2006) *Rural–Urban Migration in Developing Countries: A Survey of Theoretical Predications and Empirical Findings*, World Bank Policy Research Working Paper no. 3915, Washington, DC: World Bank, May.

Landes, David (1998) *The Wealth and Poverty of Nations: Why Some are so Rich and Some are so Poor*, New York: W.W. Norton.

Lemanski, Charlotte (2004) 'A new apartheid? The spatial implications of fear of crime in Cape Town, South Africa', *Environment and Urbanization*, 16(2): 101–112.

Lewis, W. Arthur (1954) 'Economic development with unlimited supplies of labor', *Manchester School of Economic and Social Studies*, 22(2): 139–191.

Leys, Colin (1996) *The Rise and Fall of Development Theory*, Oxford: James Currey.

Li, Zhigang and Fulong Wu (2006) 'Socioeconomic transformations in Shanghai (1990–2000): policy impacts in global-national-local contexts', *Cities*, 23(4): 250–268.

Liotta, Peter H. (2007) 'Human security and cities in the Greater Near East', in Human Security – Cities, *Human Security for an Urban Century: Local Challenges, Global Perspectives*, Vancouver: Canadian Consortium on Human Security.

Lipton, Michael (1977) *Why Poor People Stay Poor: Urban Bias in World Development*, London: Maurice Temple Smith.

Livi-Bacci, Massimo (2001) *A Concise History of World Population*, 3rd edn, Oxford: Blackwell.

Lourenco-Lindell, Ilda (2007) 'The "glocal" strategies of urban informal workers: the multiscalar agency of organised vendors in Maputo', Paper prepared for the conference on 'Informalising Economies and New Organising Strategies in Africa', Nordic Africa Institute, Uppsala, 20–22 April.

Lugalla, Joe (1995) *Crisis, Urbanization and Urban Poverty in Tanzania: A Study of Urban Poverty and Survival Politics*, Lanham, MD: University Press of America.

Lund, Frances and Caroline Skinner (2004) 'Integrating the informal economy in urban planning and governance: a case study of the process of policy development in Durban, South Africa', *International Development Planning Review*, 26(4): 431–456.

Lynch, Kenneth (2005) *Rural–Urban Interaction in the Developing World*, New York: Routledge.

Mabin, Alan (2002) 'Local government in the emerging national context', in S. Parnell, E. Pieterse, M. Swilling and D. Wooldridge (eds) *Democratising Local Government: The South African Experiment*, Cape Town: University of Cape Town Press.

Mabogunje, Akin L. (1990) 'Urban planning and the post-colonial state in Africa: a research overview', *African Studies Review*, 33(2): 121–203.

McCarney, P., M. Halfani and A. Rodriquez (1995) 'Towards an understanding of governance: the emergence of an idea and its implications for urban research in developing countries', in R. Stren and J. Bell (eds) *Urban Research in the Developing World*, Vol. 4, *Perspectives on the City*, Toronto: Centre for Urban and Community Studies, University of Toronto.

MacGaffey, Janet and Remy Banzenguissa-Ganga (2000) *Congo-Paris: Transnational Traders on the Margins of the Law*, Bloomington, IN: Indiana University Press.

McGranahan, Gordon, Deborah Balk and Bridget Anderson (2007) 'The rising tide: assessing the risks of climate change and human settlements in low-elevation coastal zones', *Environment and Urbanization*, (19)1: 17–37.

McIlwaine, Cathy and Caroline Moser (2007) 'Living in fear: how the urban poor perceive violence, fear and insecurity', in Kees Koonings and Dirk Kruijt (eds) *Fractured Cities: Social Exclusion, Urban Violence and Contested Spaces in Latin America*, London: Zed.

McIntosh, Roderick J. and Susan Keech McIntosh (1981) 'The inland Niger Delta before the Empire of Mali: evidence from Jenne-Jenno', *Journal of African History*, 22: 1–22.

Maitra, Asesh Kumar and Arvind Krishan (2000) 'Agenda 21 and urban India', in N. Low, B. Gleeson, I. Elander and R. Lidskog (eds) *Consuming Cities*, London: Routledge.

Maloney, William F. (1999) 'Does informality imply segmentation of labour markets? Evidence from sectoral transitions in Mexico', *World Bank Economic Review*, 13(2): 275–302.

Mamdani, Mahmood (1996) *Citizen and Subject: Contemporary Africa and the Legacy of Late Colonialism*, Princeton, NJ: Princeton University Press.

Mangin, William (1967) 'Latin American squatter settlements: a problem and a solution', *Latin American Research Review*, 2(3): 65–98.

Marcotullio, Peter and Gordon McGranahan (2007) 'Scaling the urban challenge', in Peter Marcotullio and Gordon McGranahan (eds) *Scaling Urban Environmental Challenges: From Local to Global and Back*, London: Earthscan.

Marcuse, Peter (1997) 'Walls of fear and walls of support', in N. Ellin (ed.) *Architecture of Fear*, New York: Princeton Architectural Press.

Marcuse, Peter (2004) 'The "War on Terrorism" and life in cities after September 11, 2001', in Stephen Graham (ed.) *Cities, War and Terrorism: Towards an Urban Geopolitics*, Oxford: Blackwell.

Marshall, Alfred (1920) *Principles of Economics: An Introductory Volume*, London: Macmillan.

Martin, Richard (1983) 'Upgrading', in R.J. Skinner and M.J. Rodell (eds) *People, Poverty and Shelter: Problems of Self-Help Housing*, London: Methuen.

Marx, Karl and Friedrich Engels (1848) *The Communist Manifesto*, London: Penguin Classics.

Meagher, Kate (1995) 'Crisis, informalization and the urban informal sector in Sub-Saharan Africa', *Development and Change*, 26(2): 259–284.

Meagher, Kate (2003) 'A back door to globalisation? Structural adjustment, globalisation and transborder trade in West Africa', *Review of African Political Economy*, 95: 57–75.

Meier, Gerald M. (1995) *Leading Issues in Economic Development*, Oxford: Oxford University Press.

Méndez, Juan, Guillermo O'Donnell and Paul Sérgio Pinheiro (eds) (1999) *The (Un)rule of Law and the Underprivileged in Latin America*, Notre Dame, IN: University of Notre Dame Press.

Mitlin, Diana (2004a) *Understanding Urban Poverty: What the Poverty Reduction Strategy Papers Tell Us*, paper prepared for the Department for International Development (DFID), London: DFID.

Mitlin, Diana (2004b) 'Civil society organizations: do they make a difference to urban poverty?', in N. Devas with P. Amis, J. Beall, U. Grant, D. Mitlin, F. Nunan and C. Rakodi, *Urban Governance, Voice and Poverty in the Developing World*, London: Earthscan.

Mitlin, Diana and David Satterthwaite (1996) 'Sustainable development and cities', in Cedric Pugh (ed.) *Sustainability, the Environment and Urbanization*, London: Earthscan.

Mitullah, Winnie (1991) 'Hawking as a survival strategy for the urban poor in Nairobi: the case of women', *Environment and Urbanization*, 3: 13–22.

Moser, Caroline (1993) *Gender Planning and Development: Theory, Practice and Training*, London: Routledge.

Moser, Caroline (1996) *Confronting Crisis: A Comparative Study of Household Responses in Four Poor Urban Communities*, Environmentally Sustainable

Development Studies and Monograph Series no. 8, Washington, DC: World Bank.
Moser, Caroline (1998) 'The asset vulnerability framework: reassessing urban poverty reduction strategies', *World Development*, 26(1): 1–19.
Moser, Caroline (2007) 'Asset accumulation policy and poverty reduction', in Caroline Moser (ed.) *Reducing Global Poverty: The Case for Asset Accumulation*, Washington, DC: Brookings Institution.
Moser, Caroline and Cathy McIlwaine (2004) *Encounters with Violence in Latin America: Urban Poor Perceptions from Colombia and Guatemala*, London: Routledge.
Moser, Caroline and Linda Peake (eds) (1987) *Women, Human Settlements and Housing*, New York: Tavistock.
Moser, Caroline and Dennis Rodgers (2005) *Change, Violence and Insecurity in Non-Conflict Situations*, ODI Working Paper 245, London: Overseas Development Institute.
Mumford, Lewis (1937) 'What is a city', *Architectural Record*, reprinted in Richard Le Gates and Fredric Stout (eds) *The City Reader*, 3rd edn, London: Routledge.
Mumford, Lewis (1961) *The City in History*, London: Martin Secker & Warburg.
Myers, Gareth Andrew (2003) *Verandahs of Power: Colonialism and Space in Urban Africa*, Syracuse, NY: Syracuse University Press.
Myrdal, Gunnar (1957) *Economic Theory and Underdeveloped Regions*, New York: Harper & Row.
Nantulya, Vinand M. (2002) 'The neglected epidemic: road traffic injuries in developing countries', *British Medical Journal*, 324(7346): 1139–1141.
Narayan, D., with R. Patel, K. Schafft, A. Rademacher and S. Koch-Schulte (2000a) *Voices of the Poor, Can Anyone Hear Us?* Oxford: Oxford University Press for the World Bank.
Narayan, D., R. Chambers, M. Shah and P. Petesch (2000b) *Voices of the Poor, Crying Out for Change*, Oxford: Oxford University Press for the World Bank.
National Public Radio (NPR) (2002) *Aids 2002 Special Report*, Washington, DC: NPR, available at www.npr.org/news/specials/aids2002/index.html (accessed 4 February 2009).
Nelson, Joan M. (1979) *Access to Power: Politics and the Urban Poor in Developing Nations*, Princeton, NJ: Princeton University Press.
North, Douglas (1990) *Institutions, Institutional Change and Economic Performance*, Cambridge: Cambridge University Press.
O'Conner, Anthony (1983) *The African City*, London: Hutchinson.
Oxfam (2007) *Adapting to Climate Change: What's Needed in Poor Countries, and Who Should Pay*, Oxfam Briefing Paper 104, Banbury: Oxfam.
Oyeniyi, Bukola (2007) 'Road Transport Workers' Union: the paradox of negotiating socio-economic and political space in Nigeria', paper prepared for the conference on 'Informalising Economies and New Organising Strategies in Africa', Nordic Africa Institute, Uppsala, 20–22 April.
Pacione, Michael (2005) *Urban Geography: A Global Perspective*, 2nd edn, London: Routledge.

Palma, Gabriel (1981) 'Dependency and development: a critical overview', in Dudley Seers (ed.) *Dependency Theory: A Critical Assessment*, London: Frances Pinter.

Pansters, Wil and Hector Castillo Berthier (2007) 'Mexico City', in Kees Koonings and Dirk Kruijt (eds) *Fractured Cities: Social Exclusion, Urban Violence and Contested Spaces in Latin America*, London: Zed.

Parker, Geoffry (2004) *Sovereign City: The City-state through History*, London: Reaktion.

Parnell, S., E. Pieterse, M. Swilling and D. Wooldridge (2002) *Democratising Local Government: The South African Experience*, Cape Town: University of Cape Town Press.

Parr, John B. (1999a) 'Growth-pole strategies in regional and economic planning: a retrospective view (Part 1: Origins and advocacy)', *Urban Studies*, 36(7): 1195–1215.

Parr, John B. (1999b) 'Growth-pole strategies in regional and economic planning: a retrospective view (Part 2: Implementation and outcome)', *Urban Studies*, 36(8): 1247–1268.

Patel, Raj (2008) *Stuffed and Starved: The Hidden Battle for the World Food System*, New York: Melville House.

Patel, Sheela, Celine d'Cruz and Sundar Burra (2002) 'Beyond evictions in a global city: people-managed resettlement in Mumbai', *Environment and Urbanisation*, 14(1): 159–172.

Patibandla, Murali and Bent Petersen (2002) 'Role of transnational corporations in the evolution of a high-tech industry: the case of India's software industry', *World Development*, 30(9): 1561–1577.

Payne, Geoffrey (1997) *Urban Land Tenure and Property Rights in Developing Countries: A Review*, London: Overseas Development Agency (ODA) and Intermediate Technology Publications.

Pearce, Jennifer (1998) 'From civil war to "civil society": has the end of the Cold War brought peace to Central America?', *International Affairs*, 74(3): 589–590.

Perry, G.E., W.F. Maloney, O.S. Arias, P. Fajnzylber, A.D. Mason and J. Saavedra-Chanduvi (2007) *Informality: Exit and Exclusion*, Washington, DC: World Bank.

Pieterse, Edgar (2008) *City Futures: Confronting the Crisis of Urban Development*, London: Zed.

Pilkington, Ed (2008) 'Eaten up', *Guardian*, 29 July.

Pinheiro, Paul Sérgio (1996) 'Democracy without citizenship', *NACLA Report on the Americas*, 30(2): 17–23.

Porter, Michael (2000) 'Location, competition, and economic development: local clusters in a global economy', *Economic Development Quarterly*, 14(1): 15–34.

Portes, Alejandro and William Haller (2005) 'The informal economy', in N.J. Smelser and R. Swedberg (eds) *The Handbook of Economic Sociology*, Princeton, NJ: Princeton University Press.

Potter, Robert (1992) *Urbanization in the Third World*, Oxford: Oxford University Press.

Potts, Deborah (2006) '"Restoring order"? Operation Murambatsvina and the urban crisis in Zimbabwe', *Journal of Southern African Studies*, 32(2): 273–291.

Potts, Deborah (2007) The State and the Informal Sector in Sub-Saharan African Economies: Revisiting Debates on Dualism', Crisis States Research Centre Working Paper no. 18 (series two), London: Crisis States Research Centre, Development Studies Institute, London School of Economics.

Pryer, Jane (2003) *Poverty and Vulnerability in Dhaka Slums*, Aldershot: Ashgate.

Pugh, Cedric (1995) 'The role of the World Bank in housing', in Brian Aldrich and Ranvinder Sandhu (eds) *Housing the Urban Poor: Policy and Practice in Developing Countries*, London: Zed.

Rakodi, Carole (1999) 'A capital assets framework for analysing household livelihood strategies', *Development Policy Review*, 17(3): 315–342.

Rakodi, Carole with Tony Lloyd-Jones (eds) (2002) *Urban Livelihoods: A People-Centred Approach to Reducing Poverty*, London: Earthscan.

Ravallion, Martin, Shaohua Chen and Prem Sangraula (2007) *New Evidence on the Urbanization of Global Poverty*, Policy Research Working Paper no. 4199, Washington, DC: World Bank, available at econ.worldbank.org/docsearch (accessed 9 June 2008).

Reader, John (2005) *Cities*, London: Vintage.

Redclift, Michael (1987) *Sustainable Development: Exploring the Contradictions*, London: Methuen.

Rees, William (1992) 'Ecological footprints and appropriated carrying capacity', *Environment and Urbanization*, 4(2): 121–130.

Reid, Hannah and David Satterthwaite (2007) 'Climate change and cities', *IIED Sustainable Development Opinion*, London: International Institute for Environment and Development.

Richards, Susan (2002) 'More trouble in paradise', *Open Democracy*, available at www.opendemocracy.net/conflict-witnessconflict/article_848.jsp (accessed 4 February 2009).

Rizzo, Matteo (2002) 'Being taken for a ride: privatisation of the Dar es Salaam transport system 1983–1998', *Journal of Modern African Studies*, 40(1): 133–157.

Robinson, Jennifer (2002) 'Global and world cities: a view from off the map,' *International Journal of Urban and Regional Research*, 26(3): 531–554.

Robinson, Jennifer (2006) *Ordinary Cities*, London: Routledge.

Rodgers, Dennis (2004) '"Disembedding" the city: crime, insecurity and spatial organization in Managua, Nicaragua', *Environment and Urbanization*, 16(2): 113–124.

Rodgers, Dennis (2005) *Unintentional Democratisation? The Argentinazo and the Politics of Participatory Budgeting in Buenos Aires, 2001–2004*, London: Crisis States Research Centre, London School of Economics.

Rodgers, Dennis (2006) 'Living in the shadow of death: gangs, violence, and social order in urban Nicaragua, 1996–2002', *Journal of Latin American Studies*, 38(2): 267–292.

Rodgers, Dennis (2007a) *Slum wars of the 21st Century: The New Geography of*

Conflict in Central America, Cities Theme Working Paper 10, London: Crisis States Research Centre, London School of Economics.

Rodgers, Dennis (2007b) 'Managua', in Kees Koonings and Dirk Kruijt (eds) *Fractured Cities: Social Exclusion, Urban Violence and Contested Spaces in Latin America*, London: Zed.

Rodrik, Dani (2003) 'Institutions, integration, and geography: in search of the deep determinants of economic growth', in Dani Rodrik (ed.) *In Search of Prosperity: Analytic Narratives on Economic Growth*, Princeton, NJ: Princeton University Press.

Rodrik, Dani, Arvind Subramanian and Franceso Trebbi (2004) 'Institutions rule: the primacy of institutions over geography and integration in economic development', *Journal of Economic Growth*, 9(2): 131–165.

Rondinelli, D.A. (1981) 'Government decentralization in comparative theory and practice in developing countries' *International Review of Administrative Science*, 47 (2), 133–147.

Rondinelli, Dennis A. (1994) 'Urbanization policy and economic growth in Sub-Saharan Africa: the private sector's role in urban development', in James D. Tarver (ed.) *Urbanization in Africa: A Handbook*, Westport, CT: Greenwood Press.

Rosenthal, Stuart and William Strange (2004) 'Evidence on the nature and sources of agglomeration economies', in J.V. Henderson and J. Thisse (eds) *Handbook of Regional and Urban Economics*, Vol. 4, Amsterdam: North-Holland.

Rostow, W.W. (1960) *The Stages of Economic Growth: A Non-communist Manifesto*, Cambridge: Cambridge University Press.

Sacquet, A. (2002) *World Atlas of Sustainable Development*, Paris: Autrement.

Sandler, Todd and Walter Enders (1999) 'Is transnational terrorism becoming more threatening? A time-series investigation', *Journal of Conflict Resolution*, 44(3): 307–332.

Sandler, Todd and Walter Enders (2005) 'Transnational terrorism: an economic analysis', in H.W. Richardson, P. Gordon and J.E. Moore II (eds) *The Economic Impact of Terrorist Attacks*, Cheltenham, UK: Edward Elgar.

Sassen, Saskia (1991) *The Global City: New York, London, and Tokyo*, Princeton, NJ: Princeton University Press.

Sassen, Saskia (2006) *Cities in a World Economy*, 3rd edn, Thousand Oaks, CA: Pine Forge Press.

Satterthwaite, David (1999) 'The key issues and the works included', in David Satterthwaite (ed.) *The Earthscan Reader in Sustainable Cities*, London: Earthscan.

Satterthwaite, David (2006a) 'Small urban centres and large villages: the habitat for much of the world's low-income population', in Cecilia Tacoli (ed.) *The Earthscan Reader in Rural–Urban Linkages*, London: Earthscan.

Satterthwaite, David (2006b) 'Towards a real-world understanding of less ecologically damaging patterns of urban development', *Environment and Urbanization*, 18(2): 1–6.

Satterthwaite, D., S. Huq, M. Pelling, H. Reid and P. Romero Lankao (2007)

Adapting to Climate Change in Urban Areas: The Possibilities and Constraints in Low- and Middle-Income Nations, IIED Human Settlements Discussion Paper Series, Theme: Climate Change and Cities no. 1, London: International Institute for Environment and Development.

Scheper-Hughes, Nancy (1992) *Death without Weeping: The Violence of Everyday Life in Brazil*, Berkeley, CA: University of California Press.

Schmidt, David (2008) 'From spheres to tiers: conceptions of local government in South Africa in the period 1994–2006', in M. van Donk, M. Swilling, E. Pieterse and S. Parnell (eds) *Consolidating Developmental Local Government, Lessons from the South African Experience*, Cape Town: University of Cape Town Press.

Schou, Arild and Marit Haug (2005) *Decentralisation in Conflict and Post-Conflict Situations*, NIBR Working Paper 2005:139, Oslo: Norwegian Institute for Urban and Regional Research (NIBR).

Schwartz, Michael (2007) 'Neo-liberalism on crack: cities under siege in Iraq', *City*, 11(1): 21–69.

Scoones, Ian (1998) *Sustainable Rural Livelihoods: A Framework for Analysis*, IDS Working Paper 72, Brighton: Institute of Development Studies, University of Sussex.

Scott, Allen J. (2001) 'Introduction', in Allen J. Scott (ed.) *Global City-Regions: Trends, Theory, Policy*, Oxford: Oxford University Press.

Scott, Allen J. and Michael Storper (2003) 'Regions, globalization, development', *Regional Studies*, 37(6–7): 579–593.

Scott, A.J., J. Agnew, E. Soja and M. Storper (2001) 'Global city-regions', in Allen J. Scott (ed.) *Global City-Regions: Trends, Theory, Policy*, Oxford: Oxford University Press.

Scott, James (1985) *Weapons of the Weak: Everyday Forms of Peasant Resistance*, New Haven, CT: Yale University Press.

Segbers, Klaus (ed.) (2007) *The Making of Global City Regions: Johannesburg, Mumbai/Bombay, Sao Paulo, and Shanghai*, Baltimore, MD: Johns Hopkins University Press.

Sen, Amartya (1981) *Poverty and Famines: An Essay on Entitlement and Deprivation*, Oxford: Clarendon Press.

Sen, Amartya (1999) *Development as Freedom*, New York: Anchor.

Sharan, Awadhendra (2006) 'In the city, out of place: environment and modernity, Delhi 1860s to 1960s', *Economic and Political Weekly*, 41(47): 4905–4911.

Sharma, Kamal (2000) *Rediscovering Dharavi: Stories from Asia's Largest Slum*, Delhi: Penguin.

Sharma, Kamal (2004) 'In a city like Mumbai', *Our Planet*, 14(4): 22–23.

SIDA (2006) *Fighting Poverty in an Urban World, Support to Urban Development*, Department for Infrastructure and Economic Cooperation, Division for Urban Development (INEC/URBAN), Stockholm: Swedish International Development Cooperation Agency.

Simone, AbdouMaliq (2004) *For the City Yet to Come: Changing African Life in Four Cities*, Durham, NC: Duke University Press.

Skinner, Caroline (2007) 'The struggle for the streets: processes of exclusion and inclusion of street traders in Durban, South Africa', paper prepared for the Living on the Margins Conference, Stellenbosch, South Africa, 26–28 March.

Smith, Adam (1970/1776) *Enquiry into the Nature and Causes of the Wealth of Nations*, Harmondsworth, UK: Penguin.

Smith, Carol A. (1995) 'Types of city-size distributions: a comparative analysis', in Ad van der Woude, Akira Hayami and Jan de Vries (eds) *Urbanization in History: A Process of Dynamic Interactions*, Oxford: Clarendon Press.

Soja, Edward W. (2000) *Postmetropolis: Critical Studies of Cities and Regions*, Oxford: Blackwell.

Southall, A. (1988) 'Small towns in Africa revisited', *African Studies Review*, 31(3): 84–97.

Souza, Celine (2001) 'Participatory budgeting in Brazilian cities: limits and possibilities in building democratic institutions', *Environment and Urbanization*, 13(1): 159–184.

Sovani, N.V. (1964) 'The analysis of "over-urbanization",' *Economic Development and Cultural Change*, 12(2): 113–122.

Stanislawski, Dan (1947) 'Early Spanish town planning in the New World', *Geographical Review*, 37(1): 94–105.

Staudt, Kathleen (2001) 'Informality knows no borders? Perspectives from El Paso-Juárez', *SAIS Review*, 21(1): 123–130.

Stephens, Caroline (1996) 'Healthy cities or unhealthy islands: the health and social implications of urban inequality', *Environment and Urbanization*, 8(2): 9–30.

Stewart, Frances (1985) *Basic Needs in Developing Countries*, Baltimore, MD: Johns Hopkins University Press.

Stiglitz, Joseph (2002) *Globalization and its Discontents*, New York: W.W. Norton.

Stoker, Gerry (1998) 'Public–private partnerships and urban governance', in Jon Pierre (ed.) *Partnerships in Urban Governance: European and American Experience*, London: Macmillan.

Storper, Michael and Anthony J. Venables (2004) 'Buzz: face-to-face contact and the urban economy,' *Journal of Economic Geography*, 4: 351–370.

Streeten, P. with S. Burki, M. Haq, N. Hicks and F. Stewart (1981) *First Things First: Meeting Basic Human Needs in Developing Countries*, Oxford: Oxford University Press.

Tacoli, Cecilia (1998) 'Rural–urban interactions: a guide to the literature', *Environment and Urbanization*, 10(1): 147–166.

Tannerfeldt, Göran and Pers Ljung (2006) *More Urban Less Poor: An Introduction to Urban Development and Management*, London: Earthscan.

Taschner, Suzana P. (1995) 'Squatter settlements and slums in Brazil: twenty years of research and policy', in Brian Aldrich and Ravinda Sandhu (eds) *Housing the Urban Poor: Policy and Practice in Developing Countries*, London: Zed.

Taylor, Peter (2004) *World City Network: A Global Urban Analysis*, London: Routledge.

Tendler, Judith (1997) *Good Government in the Tropics*, Baltimore, MD: Johns Hopkins University Press.

Tendler, Judith (2000) 'Why are social funds so popular?', in Shahid Yusef, Weiping Wu and Simon Everett (eds) *Local Dynamics in the Era of Globalization*, companion volume of *World Development Report 1999/2000*, New York: World Bank and Oxford University Press.

Tendler, Judith and Sara Freedheim (1994) 'Trust in a rent-seeking world: health and government transformed in Northeast Brazil', *World Development*, 22(12): 1771–1791.

Thomas, Alan (2000a) 'Poverty and the "end of development"', in Tim Allen and Alan Thomas (eds) *Poverty and Development in the 21st Century*, Oxford: Oxford University Press.

Thomas, Alan (2000b) 'Meanings and views of development', in Tim Allen and Alan Thomas (eds) *Poverty and Development in the 21st Century*, Oxford: Oxford University Press.

Tibaijuka, Anna Kajumulo (2000) 'A revitalized habitat committed to fighting the urbanization of poverty', *United Nations Chronicle*, 37(4), available at www.un.org/Pubs/chronicle/2000/issue4/0400p54.htm (accessed 4 February 2009).

Tilly, Charles (1994) 'Entanglements of European cities and states', in *Cities and the Rise of States in Europe: A.D. 1000 to 1800*, Boulder, CO: Westview Press.

Todaro, Michael (2000) *Economic Development*, 7th edn, Harlow, UK: Pearson Education.

Townsend, Peter (1993) *The International Analysis of Poverty*, Hemel Hempstead: Harvester Wheatsheaf.

Toye, John (1987) *Dilemmas of Development*, Oxford: Blackwell.

Turner, John (1972) 'Housing as a verb', in John Turner and Robert Fichter (eds) *Freedom to Build*, London: Macmillan.

Turner, John (1976) *Housing by People: Towards Autonomy in Building*, London: Marion Boyars.

UNCHS (1996) *An Urbanizing World: Global Report on Human Settlements*, Oxford: Oxford University Press.

UNCHS (2001a) *The Right to Adequate Housing: A Major Commitment of the Habitat Agenda*, Nairobi: United Nations Centre for Human Settlements (UN-Habitat).

UNCHS (2001b) *The State of the World's Cities 2001*, Nairobi: United Nations Centre for Human Settlements (UN-Habitat).

UNDP (1994) *Human Development Report 1994*, New York: Oxford University Press.

UNDP (1997) *Governance for Sustainable Human Development*, New York: United Nations Development Programme.

UNDP (2003) *Human Development Report 2003*, New York: United Nations Development Programme.

UNDP (2007) *Human Development Report 2007/2008*, New York: United Nations Development Programme.

UN-Habitat (2003a) *Water and Sanitation in the World's Cities*, London: UN-Habitat and Earthscan.

UN-Habitat (2003b) *The Challenge of Slums: Global Report on Human Settlements 2003*, London: Earthscan.

UN-Habitat (2003c) *The Habitat Agenda, Goals and Principles, Commitments and the Global Plan of Action*, Nairobi: UN-Habitat, available at www.unhabitat.org/downloads/docs/1176_6455_The_Habitat_Agenda.pdf (accessed 4 February 2009).

UN-Habitat (2005) *Report of the Fact-Finding Mission to Zimbabwe to Assess the Scope and Impact of Operation Murambatsvina by the UN Special Envoy on Human Settlements Issues in Zimbabwe*, available at www.reliefweb.int/rw/rwb.nsf/db900SID/HMYT-6EJM2G?OpenDocument (accessed 4 February 2009).

UN-Habitat (2006) *State of the World's Cities Report 2006/7*, Nairobi: United Nations Centre for Human Settlements.

UN-Habitat (2007) *Global Report on Human Settlements 2007: Enhancing Urban Safety Security*, London: Earthscan.

United Nations (Department of Economic and Social Affairs) (1999) *The World at Six Billion*, New York: United Nations.

United Nations (World Water Assessment Programme) (2003) *Water for People, Water for Life: The United Nations World Water Development Report 2003*, Barcelona: UNESCO and Berghan.

United Nations (Department of Economic and Social Affairs/Population Division) (2004) *World Urbanization Prospects: The 2003 Revision*, New York: United Nations.

United Nations (Department of Economic and Social Affairs/Population Division) (2007) *World Population Policies 2007*, New York: United Nations.

United Nations (Department of Economic and Social Affairs/Population Division) (2008) *World Urbanization Prospects: The 2007 Revision*, New York: United Nations.

Van Brabant, Koenraad (2007) 'Human insecurity in six post-conflict cities', in Human Security – Cities, *Human Security for an Urban Century: Local Challenges, Global Perspectives*, Vancouver: Canadian Consortium on Human Security.

Van Dijk, Jan, John van Kesteren and Paul Smit (2007) *Crime Victimisation in International Perspective: Key Findings from the 2004–2005 ICVS and EU ICS*, The Hague: Boom Legal Publishers.

Van Dijk, Meine Pieter (2003) 'Government policies with respect to an information technology cluster in Bangalore, India', *European Journal of Development Research*, 15(2): 93–108.

Van Donk, Mirjam (2006) '"Positive" urban futures in Sub-Saharan Africa: HIV/AIDS and the need for ABC (A Broader Conceptualisation)', *Environment and Urbanization*, 18(1): 155–175.

Van Horen, Basil (2004) 'Fragmented coherence: solid waste management in Colombo', *International Journal of Urban and Regional Research*, 28(4): 757–773.

Vargas, Lucinda (1999) 'The binational importance of the Maquiladora Industry', *Southwest Economy*, 6 (November–December), Federal Reserve Bank of Dallas: 1–5.

Vargas, Lucinda (2001) 'Maquiladoras: impact on Texas border cities', *The Border Economy*, Federal Reserve Bank of Dallas, June.

Varshney, Ashutosh (2002) *Ethnic Conflict and Civic Life: Hindus and Muslims in India*, New Haven, CT: Yale University Press.

Wade, Robert (2004, originally published 1990) *Governing the Market: Economic Theory and the Role of Government in East Asian Industrialization*, 2nd edn, Princeton, NJ: Princeton University Press.

Walton, John (1998) 'Urban conflict and social movements in poor countries: theory and evidence of collective action', *International Journal of Urban and Regional Research*, 22(3): 460–481.

Warwick, Hugh and Vicky Cann (eds) (2007) *Going Public, Southern Solutions to the Global Water Crisis*, London: World Development Movement.

WaterAid (2001) *Looking Back: The Long-Term Impacts of Water and Sanitation Projects*, London: WaterAid.

Watt, Paul (2000) *Social Investment in Economic Growth: A Strategy to Eradicate Poverty*, Oxford: Oxfam.

White, Gordon (1995) 'Towards a democratic developmental state', *IDS Bulletin*, 26(2): 27–36.

WHO (2002) 'Mumbai slum dwellers sewage project goes nationwide', *Bulletin of the World Health Organization*, 80(8): 684–685, available at whqlibdoc.who.int/bulletin/2002/Vol80-No8/bulletin_2002_80(8)_684-687.pdf (accessed 4 February 2009).

WHO (2004) 'World Health Day: road safety is no accident!', available at www.who.int/mediacentre/news/releases/2004/pr24/en (accessed 4 February 2009).

WHO/Unicef (2000) *Global Water Supply and Sanitation Assessment 2000 Report*, Geneva: World Health Organization and United Nations Children's Fund.

Willis, Katie (2005) *Theories and Practices of Development*, London: Routledge.

Wirth, Louis (1938) 'Urbanism as a way of life', *American Journal of Sociology*, 44(1): 1–24.

Wolf, Eric (1969) *Peasant Wars of the Twentieth Century*, New York: Harper & Row.

Wood, Geoff (2003) 'Staying secure, staying poor: the Faustian bargain', *World Development*, 31(3): 455–471.

World Bank (1990) *World Development Report 1990*, Washington, DC: World Bank.

World Bank (1991) *Urban Policy and Economic Development: An Agenda for the 1990s*, Washington, DC: World Bank.

World Bank (1997) *World Development Report: The State in a Changing World*, Oxford: Oxford University Press.

World Bank (2000) *World Development Report 2000*, Washington, DC: World Bank.

World Bank (2001) *Upgrading Urban Communities: Cities without Slums*, Washington, DC: World Bank.

World Bank (2005) *World Development Report 2005: A Better Investment Climate for Everyone*, Washington, DC: International Bank for Reconstruction and Development and the World Bank.

World Bank (2006) *World Development Indicators 2006*, Washington, DC: International Bank for Reconstruction and Development and World Bank.

World Commission on Environment and Development (WCED) (1988) *Our Common Future* (Brundtland Commission), Oxford: Oxford University Press.

World Development Movement (WDM) (2005) *Dirty Aid, Dirty Water: The UK Government's Push to Privatise Water and Sanitation in Poor Countries*, London: WDM.

Wratten, Ellen (1995) 'Conceptualising urban poverty', *Environment and Urbanization*, 7(1): 11–38.

Wu, Fulong (2000) 'The global and local dimensions of place-making: remaking Shanghai as a world city', *Urban Studies*, 37(8): 1359–1377.

Index